Fingerprints

Fingerprints

The Origins of Crime Detection

and the Murder Case

that Launched Forensic Science

COLIN BEAVAN

HYPERION

New York

Library of Congress Cataloging-in-Publication Data

Beavan, Colin.
  Fingerprints : the origins of crime detection and the murder case that
launched forensic science / by Colin Beavan.
     p. cm.
Includes bibliographical references and index.
ISBN 0-7868-6607-1
  1. Fingerprints. 2. Fingerprints—Identification. 3. Criminal investigation.
4. Murder—Investigation—Great Britain. I. Title.

HV6074.B34 2001
363.25'8—dc21

                                                        00-066365

Paperback ISBN 0-7868-8528-9

Original hardcover design by Caroline Cunningham

FIRST PAPERBACK EDITION

10   9   8   7   6   5   4   3   2   1

To my Mom and Dad

Acknowledgments

This project began in the office of my agent, Eric Simonoff. We were batting around ideas, when Eric suddenly said that he'd always thought the history of fingerprints might make an interesting story. He suggested I look into it. I looked, found the fascinating tale of Henry Faulds, and this book resulted.

Experiencing the amazing kindness of people like Eric is part of what has made this book such a pleasure to write. Peternelle van Arsdale, my editor, always knew the exact question to ask and just how to ask it without sending me spiraling into paroxysms of self-doubt. She gently knocked this manuscript into shape, and knew just when to indulge me in a little joking around. Thanks, too, to Jennifer Barth, Peternelle's predecessor; to Alison Lowenstein, her assistant editor; to Susan Groarke, the copy editor; and to Hyperion's design, production, and publicity staff.

Keith Beavan and Michelle Conlin have read and reread this book, patiently making suggestions and encouraging me along the way. Samuel and Hugo Douglas Freeman, two of the world's loveliest little boys, and their mother and father, Sarah and Michael, gave me many days of shelter while I did research in London. Liz and Will Pemble put me up in San Francisco.

Henry Faulds's only surviving relatives—his great-nephew, Robert Stewart, and great-niece, Catherine Stewart—both kindly invited me into their homes and let me rummage through old family papers. John Berry, a retired fingerprint examiner and former editor of *Fingerprint Whorld,* and his wife, fed me, entertained me with stories, provided me with documents, and taught me how fingerprints work. John also provided two diagrams for this book. Martin Leadbetter, head of the Cambridgeshire Police Fingerprint Branch, very kindly helped me with research and technical detail. Pat Wertheim, an American fingerprint consultant, opened many doors and had numerous technical conversations with me. Maurice Garvey, Fabienne Smith, and Anne Joseph sent me papers on fingerprint history.

Ken Aoki translated material about Faulds from Japanese. Tony Simpson at John Jay College and Lin Salmo at the Mark Twain Project kindly helped with work at their libraries. Enormous thanks also to the numerous librarians who guided me around the collections of the New York Public Library, The British National Library, The Scottish National Library, The Bancroft Library at the University of California, The Library of the University of Texas at Austin, the Galton Archives at University College London, and The FBI Academy Library. Finally, heartfelt thanks to friends and family who have supported me so well.

Every human being carries with him from his cradle to his grave certain physical marks which do not change their character, and by which he can always be identified—and that without shade of doubt or question. These marks are his signature, his physiological autograph, so to speak, and this autograph can not be counterfeited, nor can he disguise it or hide it away, nor can it become illegible by the wear and mutations of time. This signature is not his face—age can change that beyond recognition; it is not his hair, for that can fall out; it is not his height, for duplicates of that exist; it is not his form, for duplicates of that exist also, whereas this signature is this man's very own—there is no duplicate of it among the swarming populations of the globe. . . . This autograph consists of the delicate lines or corrugations with which Nature marks the insides of the hands and the soles of the feet.

—Samuel Clemens, writing as
Mark Twain, in *The Tragedy of
Pudd'nhead Wilson*, 1894

Contents

Chronology of Fingerprints

400 A.D.   Anglo-Saxons conjure "evidence" of criminal guilt during supernatural ordeals.

1215 A.D.   Pope Innocent III forbids clergy from taking part in ordeals. System of investigating juries begins in England.

1504 A.D.   Henry VII ratifies first law calling for eyewitness testimony. The word "evidence" is introduced into English law.

1764 A.D.   In Italy, Cesare Beccaria publishes *Dei deletti e delle pene (Crimes and Punishment)*, heralding the rationalization of law and the burgeoning of prisons.

1812 A.D.   In France, François-Eugène Vidocq establishes Europe's first official detective branch and pioneers the use of physical evidence.

1816 A.D.   Britain opens first national penitentiary at Millbank.

1842 A.D.   Vidocq-style detective force established in Britain.

1858 A.D.   William Herschel begins privately experimenting with fingerprints in India.

1863 A.D.   Garroting epidemic scares public that hordes of criminals, once dispatched by the hangman or deportation, now roam the streets of London.

1869 A.D.   Habitual Criminals Act in England provides longer sentences for hardened criminals with previous convictions. Need to identify prior offenders first arises in Britain.

1870 A.D. "The Claimant" sues for the title of Baronet of Tichborne, falsely identifying himself as the true heir, who was lost at sea fifteen years earlier. This case eventually sparks fingerprint concept in Dr. Henry Faulds's mind.

1877 A.D. Herschel, still in India, begins year-long use of fingerprints as signatures on land titles and jailers' warrants.

1878 A.D. Faulds, a Scottish missionary working in Japan, discovers fingerprints on ancient pottery and begins extensive experiments.

1880 A.D. Faulds becomes first person to publicly suggest fingerprints as a method of criminal identification in a letter published in *Nature*.

1883 A.D. Alphonse Bertillon, in Paris, identifies his first habitual criminal using his newly installed anthropometric system of measurements.

1886 A.D. Henry Faulds begins trying to convince Scotland Yard to adopt fingerprints.

1888 A.D. Francis Galton begins experimenting with fingerprints as a means of determining physical and intellectual prowess.

1892 A.D. In Argentina, fingerprint evidence leverages a confession from a mother who murdered her two children, though news of the case does not reach Europe for many years.

1893 A.D. Edward Henry, chief of police in Bengal, India, adds thumbprints to the anthropometric records he began taking the previous year.

1894 A.D. Britain adopts an identification system which is a hybrid of anthropometry and fingerprints.

1896 A.D. Adolf Beck, an innocent man, is jailed for five years after being wrongly recognized as a known con artist by police and a witness. Fingerprints would have shown he was the wrong man.

1897 A.D. Henry's assistant Azizul Haque comes up with a comprehensive system for classifying fingerprints, making practical their use without measurements.

1901 A.D. Britain adopts the fingerprint classification system developed largely by Haque, but which has come to be known as the "Henry classification system."

1902 A.D. Harry Jackson found guilty of burglary on evidence of fingerprints. First time fingerprints used to prove guilt in a British courtroom.

1904 A.D. United States Bureau of Identification establishes fingerprint collection.

1905 A.D. The Stratton brothers are tried and hanged on fingerprint evidence for the vicious murder of Thomas and Ann Farrow. Henry Faulds takes their side against police.

1911 A.D. Thomas Jennings is the first to be convicted of murder in the United States on the basis of fingerprint evidence.

1911 A.D. Galton dies.

1913 A.D. Bertillon dies.

1917 A.D. Herschel dies.

1930 A.D. Faulds dies.

1938 A.D. Scottish judge George Wilton begins campaign for Faulds's recognition as a fingerprint pioneer.

1987 A.D. American fingerprint experts restore Faulds's grave.

1999 A.D. Federal Bureau of Investigation installs massive fingerprint
           computer capable of storing the fingerprints of 65 million
           individuals.

Fingerprints

One

The Shocking Tragedy
at Deptford

Most mornings, young William Jones burst through the unlocked door of Chapman's Oil and Colour Shop, heard the tinkle of the bell, and breathed in the sharp-smelling air, heavy with the odor of paint. But today the entrance off the High Street of Deptford, near London, was locked. The sixteen-year-old pressed his shoulder against the door and shoved. No use. It wouldn't budge.

William had never known his boss, Thomas Farrow, to open the shop later than 7:30 A.M. Farrow was promoted to manager and moved into the shop's upstairs apartment with his wife in 1902. In the three years since, he'd never failed to throw open the shop-front shutters when the first early-bird customer knocked, often at sunup. But this morning's knocks had gone unanswered. Disappointed, without their supplies, the house painters rolled their wagons back into the morning throng that bustled its way to work.

By the time William arrived, the fast-walking commuters were gone and the High Street was quiet again. Only stragglers still hurried past: red-eyed butchers rushing to Deptford's slaughterhouses, unshaven sailors running to their ships moored here on the south side of the Thames. The morning rush hour was over. It was 8:35. Still, the shop door was closed. William banged at the door. Over the clip-clop of passing horse-drawn carriages, he shouted at the upstairs windows. No reply. Something was wrong.

At first, William thought the Farrows were ill. At seventy-one and sixty-five, Thomas and Ann were getting frail. But then a niggling thought reminded William that today, Monday, was banking day, the one time when Mr. Farrow's cash box swelled with a whole week's earnings. When William looked through the letter box in the door, he saw that at the far end of the shop, in the Farrows' small parlor, a large lounge chair lay tipped on its side—a bad sign. William sprinted to George Chapman's other shop in Greenwich, recruited the help of Louis Kidman, Chapman's assistant, and both ran back to Deptford. They would have to break their way in.

The boys burst through a shop adjoining Chapman's. Out back, they scaled a dividing wall and dropped into the Farrows' yard. They found the scullery door open, walked through it, and were immediately horrified by what they saw. Under the overturned lounge chair, Thomas Farrow's body lay face down, crumpled in a grotesque, bloody pile. His bald head, resting on a metal fender surrounding the fireplace hearth, had been smashed open. A pool of dark blood filled the hearth and ran into the cold ashes. The whole grisly scene shocked William so badly that he never stopped to consider that, elsewhere in the building, Mrs. Farrow might be desperately in need of help.

· · ·

Chapman's Oil and Colour Shop stood squat and two-story on the High Street, lined up shoulder to shoulder in a long row of stores. At the foot of the High Street the railway ended, the terminus for the commuter line that carried the region's clerks and

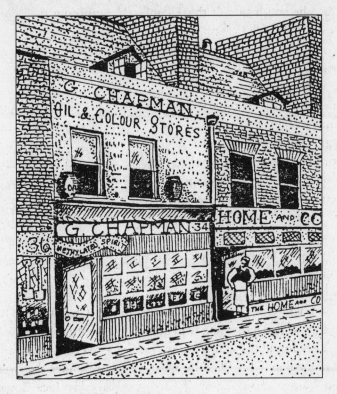

Chapman's Oil and Colour Shop

laborers to the center of London, away from the ugly town of Deptford, which for hundreds of years had been burdened with the stinking, disease-ridden industries that the capital turned away.

In the early seventeenth century, the filth-producing slaughterhouses, exiled by the capital's city fathers, had moved here. When in 1897, the world's then-largest proposed electricity-generating plant was refused a home in London, it too landed in Deptford, along with the smoke and the dirt it belched from its chimney. Only the poorest and most desperate people wound up living among this filth. Some Deptford neighborhoods were so dangerous that policemen refused to patrol them alone.

Cold-blooded murder, however, was an uncommon spectacle. That was why, when Sergeant Albert Atkinson arrived at the scene of what the papers would call "The Shocking Tragedy at Deptford," he was greeted by a rabble of curious onlookers. Inside, the sergeant, with Louis Kidman behind him, stealthily mounted the stairs to the second floor, looking for intruders who might still be hiding. What they found instead was the unconscious Mrs. Farrow. Her head was so badly smashed that both men assumed she was dead. Her moans shocked them into the realization that she still clung to life. Sergeant Atkinson hurriedly rang for the police surgeon.

Dr. Dudley Burnie arrived half an hour later, around 9:45, together with detectives from the nearby Blackheath Road police station. By that time, a large group of constables wrestled back a horde of onlookers who had been helping themselves to macabre souvenirs from inside the shop. Burnie and the policemen had to force their way through.

Burnie rushed upstairs and was in the middle of dressing the gaping, bloody wounds on Mrs. Farrow's head, when she suddenly gained consciousness and struggled violently against him, "evidently being in the state of very great fright," he later said. Ambulance men arrived and heaved Mrs. Farrow onto a stretcher to carry her down the stairs. Thomas Farrow's empty cash box lay on the floor threatening to trip them. Sergeant Atkinson picked it up and shoved it under the bed. He used his bare hands.

Dr. Burnie followed the ambulance men down the stairs and began his examination of Mr. Farrow's body. The time of death, he determined, had been between one and one and a half hours earlier. One large wound gaped open over Farrow's right eyebrow and another on the right side of his nose. The old man had also sustained two gashes above the left ear and one above the right. Later, during the autopsy, the doctor would discover that Farrow's skull had been shattered into several pieces in the region of the temple and that the right cheekbone was fractured. All in all, the police surgeon believed, Thomas Farrow had received six blows to the head, probably with a crowbar.

. . .

Scotland Yard's crack homicide detective, Chief Inspector Frederick Fox, joined the local police at the Oil and Colour Shop around 11:30 A.M. His two crime scene photographers immediately began setting up their bulky boxes and tripods. With Fox also came his boss, Scotland Yard's Assistant Commissioner Melville Macnaghten. A short, mustached man whose fastidious grooming and upright stature gave him the air of a landed gentleman, Macnaghten was in charge of the entire Criminal Investigation Department (the CID). From now on, he would call the shots in the investigation of this high-profile case.

What little his subordinates had pieced together was based on the lack of any signs of forced entry, the fact that Farrow was found still dressed in a nightshirt, and the placement of two puddles of blood. The criminals, the investigators surmised, knocked early in the morning, waking the unsuspecting old man, and telling him through the latched door that they needed painting supplies. Once inside, as Farrow busied himself with attending to their supposed requirements, they clobbered the back of his head, accounting for the puddle of blood behind the counter.

The robbers searched the shop and back parlor for the cash box but, finding nothing, started up the stairs. Farrow, back on his feet, threw himself at the invaders, desperately fighting to keep them off the second floor where his wife lay unprotected in bed. The robbers' further bone-crushing blows to Farrow's head left him bleeding at the foot of the stairs—the location of the second pool of blood. Upstairs, a few merciless swipes silenced Mrs. Farrow's screaming and, searching the bedroom, the robbers found and emptied the cash box.

Descending, they were confronted again by Mr. Farrow, miraculously revived a second time. They scuffled, overturning the furniture in the parlor, until Farrow was again struck down. The burglars rinsed Farrow's blood from their hands in a basin the police later found filled with pinkish water. They cut holes in stockings to make masks, but abandoned them when they realized they would merely attract attention. Instead, the criminals slipped out of the shop and into the rush-hour crowd, as though they were merely customers. They left behind two mortally wounded people—all for the sake of less than ten pounds (about fifty dollars then, or less than a thousand dollars in the year 2000).

Police found no eyewitnesses and no murder weapons. The lack of forced entry meant there was no way to identify the criminals from knowledge of their methods. Not even the stocking masks, found in the shop, offered any helpful clues. If the robbers had brought the masks with them, a visit to the neighborhood stocking shop might have developed a promising lead, but the thieves had cut them from Mrs. Farrow's own hose.

To make matters worse, the burglars, interested only in the shopkeeper's cash, had stolen no trinkets of value. No piece of stolen jewelry or silverware would link them to the case. Visits to pawnshops and known receivers, standard procedure after such a crime, would turn up nothing. The only hope was that Mrs. Farrow might recover consciousness and identify her attackers to the constable who waited beside her hospital bed. But the longer she remained unconscious, the greater the risk that she would succumb to a fatal bout of pneumonia.

Macnaghten saw that the case was troublesome, and it

Melville Macnaghten

brought back bad memories. Only three days after he first joined the Yard in 1889, two telegrams arrived reporting that the body parts of a woman had been found on the banks of the Thames. Macnaghten and a group of detectives slipped and slid their way along the muddy riverbank, picking over debris in search of the rest of the torso. Because the head was never found, it took a scar on the wrist to identify the body as belonging to a woman who had been reported missing from her lodging house in Chelsea. There the trail went cold. Macnaghten's first case, known as the Thames Mystery, was never solved.

This was just one year after Jack the Ripper's seven brutal murders had gone unsolved, and Macnaghten had then experienced firsthand the public's anger toward the police when they did their job badly. "Jack" had made police investigators look foolish by delivering, right under their noses, signed notes to newspapers. On Macnaghten's first day, his new boss, the CID's then Chief Constable, commenting on the public's angry response, said, "Well, my boy, you are coming into a funny place. They'll blame you if you do your duty, and they'll blame you if you don't."

Throughout the rest of his career, Macnaghten kept on his desk gruesome photographs of the Ripper's victims, each goading him to let no other cases go unsolved. He'd experienced failure, and he couldn't bear to experience it again. But because of the lack of clues turned up by his subordinates in the Farrow case, he might have no choice. This wasn't going to look good for the police.

Farrow had been killed in his own store, on the busy High Street, while commuters rushed past outside, and no policeman had seen or noticed anything suspicious. "Certain members of the police force are lacking in discernment and intelligence," the *Kentish Mercury* said of the Farrow murder. Recently, another shopkeeper's murder in Brixton had gone unsolved. Together, the crimes suggested a trend, and crime trends were what Macnaghten's CID was supposed to prevent.

Macnaghten marched purposefully past the shop counter and mounted the stairs, determined to search for clues of his own. He surveyed the bloodied chaos in the bedroom, where his eyes fell upon the cash box and its tray, jutting out from under Mrs. Farrow's bed. He carefully picked them up, but not before taking his handkerchief from his pocket to prevent his bare fingers from touching their surfaces.

For some time, a small group of officers in Macnaghten's department had been developing a technique that they claimed could identify a man from a surface that he had touched. Macnaghten and his senior colleagues envisioned a system with assembly-line efficiency, spitting out proofs of the presence of suspects at crime scenes, and closing cases that might otherwise go the way of the Ripper and Thames Mystery murders.

But the use of forensics was uncommon in these times when the boundary between science and quackery was blurry. Acceptance of the new technique depended on convincing the police ranks that it was practical, and the judiciary and the public that it was just. Macnaghten had been hoping for some time for a headline-grabbing case that might prove these points. He thought the Farrow case might be the one.

The Assistant Commissioner turned the cash box and tray over in his hands, scrutinizing the surfaces for the clue he needed. Suddenly he looked up. "Have all the men assemble up here," he told a nearby constable. Once the men had hauled themselves up the steep, narrow stairs, Macnaghten eyed them as a group. "Has anyone touched this cash box or its tray?" No one stepped forward. Macnaghten's tone and demeanor suggested that he might not be pleased with any man who answered yes. He told them to think carefully. This was important.

The officers shuffled nervously. If one had touched it and sought to hide his unintended misdeed, would Macnaghten later hear of the fact from another officer? Better to own up now than get caught in a lie. Sergeant Atkinson stepped forward. He had pushed it a little ways under the bed, he said, to ensure that the stretcher bearers didn't trip on it when they took Mrs. Farrow away. Macnaghten nodded. Again with his handkerchief, he picked up the tray of the cash box, turning its underside toward the men. On the shiny, enameled surface they saw a dull, oval smudge, such as their children's greasy fingers might leave behind after draining a glass of milk. Some understood its import, others did not.

Macnaghten handed the tray along with his handkerchief to one of his officers. Wrap it carefully in paper and make sure no one else touches it, he said. Then Macnaghten turned to Sergeant Atkinson, who was still standing in front of the assembly. Macnaghten could see that the young sergeant was embarrassed. No harm done, Macnaghten said. But he ordered the sergeant to report to Detective-Inspector Charles Collins at the

Yard so he could be sure the mark on the tray hadn't come from his fingers.

When Macnaghten left the room, those officers who understood the significance of the smudge carped among themselves. The old guard were highly suspicious of this new "scientific palmistry" that so intrigued the boss. Use of these newfangled fingerprints in such a high-profile murder investigation could bring ridicule on the Yard. Only once before had a crime-scene fingerprint been accepted in a British court, and that was just for burglary. This was murder. What jury would be willing to send a man to the gallows on the evidence of a gob of sweat smeared on a piece of metal?

.   .   .

Thomas Farrow's body was barely cold when Detective-Inspector Collins received the cash-box tray later that day. Collins was second in command of the new fingerprint branch, a part of Macnaghten's CID. Before the 1901 formation of the branch, Collins had, for many years, been forced to use old-fashioned methods of criminal identification, based on measuring bodies, photographing faces, and writing down distinguishing features. These methods were far from reliable. Now, for the first time in his career, Collins had encountered an identification system that actually worked, and he was obsessed with it. He would ultimately dedicate more than twenty-five years of his life to improving and applying the fingerprint technique.

In his office loomed a huge wooden cabinet with 1,024 pigeonholes accommodating each of the classifications into which an individual's set of ten fingerprints could fall. A handful of fingerprint experts bustled back and forth between their workbenches and the cabinet's cataloged fingertip impressions. Examined closely, a fingertip reveals a pattern of parallel ridges

A right thumbprint

interspersed with furrows, as though of a diminutive farm field. The furrows are like gutters into which moisture flows so that it is not trapped in a slippery film between the fingertip and whatever it is trying to grip.

It is not the ridges' function that makes them of interest to the identification expert, however. What fascinates him instead is the fact that the intricate ridge patterns are unique to each finger. A fingerprint expert can tell apart the marks of two digits more easily than he can differentiate two people's faces. The facial features of identical twins, for example, can be mistaken, but their fingerprints can never be confused by a trained expert, however. A person's fingerprint set is therefore a permanent and unmistakable record of his identity. It is like a biological seal which, once impressed, can never be denied. Eighty thousand such biological seals of convicted criminals crowded the pigeonholes in Scotland Yard's fingerprint branch.

This massive collection of fingerprints, however, had never before been used to collar a murderer. Sleuthing was not the fingerprint expert's primary function. Instead, Collins and his colleagues passed their days filing fingerprints taken from recent convicts, and using the previously filed fingerprints to double-check the identities of the newly arrested. Their main goal was to identify "recidivist" or "habitual" offenders who pretended to be first-timers, adopting pseudonyms in hopes of hiding their previous convictions and getting lighter sentences.

The practice of correlating a criminal's sentence with the number of his prior convictions began in the nineteenth century, when jail cells and prison guards first took the place of gallows and their hangmen. To the essentially honest man who fell on hard times and stole to feed his family, the new prison system prescribed a short stay behind bars, just enough unpleasantness to deter further crime. It was believed that the habitual offender, on the other hand, could not so easily have his criminal bent punished out of him. Long-term removal from society was thought to be the only way to prevent his misdeeds. There was one problem with this two-pronged penal approach: How do you tell the hardened criminals from the first-timers?

The first suggested use of fingerprints as a method of criminal identification came in an October 1880 issue of the prestigious scientific journal *Nature*. An article, penned by an unknown Scottish medical missionary working in Japan named Henry Faulds, proposed many of the elements of the fingerprint system as it eventually came to be used. Faulds, having studied thousands of fingerprints, would spend the next ten years trying to convince Scotland Yard to adopt the ideas in his article. The Yard dismissed Faulds as a crank, and cruelly, when it finally did adopt fingerprinting, denied that Faulds had any part in the system's conception.

A month after the publication of Faulds's article, a second article on fingerprints appeared, also in *Nature*. William Herschel, a British magistrate based in Bengal, replying to Faulds, wrote that he had used fingerprints officially as "sign-manuals," or signatures, sanctioning the idea's practicality. Still, the British establishment paid no attention to fingerprinting until, in 1888, the interest of the well-known scientist Francis Galton gave it credibility. A cousin of Charles Darwin, Galton's passion was the improvement of the human race by artificial selection. He took to fingerprints, thinking their intricate ridge patterns might somehow reveal their owners' physical and mental capacities— their worth as breeding stock.

Galton's published work sparked the interest of the Inspector-General of Police in Bengal, India, Edward Henry, who made the leap from theory to practice and applied fingerprints to police work. Henry and his assistant, Azizul Hague, developed a classification system that allowed fingerprint sets to be logically filed according to the form of their ridge patterns. Without the system, an inspector searching for a particular fingerprint set would have to rummage through the entire collection. With it, he easily went straight to the place where the set was filed. What came to be known as the Henry classification system made possible the use of fingerprint registers numbering in the many thousands, a prerequisite for practical use in criminal identification.

When widespread use of fingerprint identification proved successful in India, Henry was in 1901 recalled to London, made Assistant Commissioner of the CID, and charged with establishing Scotland Yard's new Fingerprint Branch. The branch had immediate success, cracking the pseudonyms of 632 repeat offenders in its first year. In 1905, Henry was promoted to Commissioner of Scotland Yard. He left the Fingerprint Branch in the hands of Detective-Inspector Charles Steadman and his deputy, Detective-Inspector Collins, the officer to whom Macnaghten delivered the Farrow murder cash box.

At his workbench, Collins examined the cash-box tray under his magnifying glass. Fingerprints can be impressed in anything from paint to blood, but this one, like most found at crime scenes, had been left in sweat. On the gripping surfaces of the hands and feet, 3,000 sweat glands per square inch crowd together more densely than anywhere else on the body. Keeping the skin lubricated so it does not crack, the glands also make each finger like a self-inking rubber stamp, leaving calling cards on every surface it touches.

Because of this, since most human action involves touching, fingerprints invisibly populate the world's surfaces. Taken together, these fingerprints are like pages from the Recording Angel's book of deeds, and Charles Collins, with his magnifying glass, could read them. If a fingerprint he found on an object matched a fingerprint in his cabinet, Collins could deduce the name of the person who touched the object. This is how Collins hoped to discover Farrow's murderer.

The impression on the cash-box tray followed an arch pattern and came from a right thumb. Collins could tell that it was a thumb because the impression was too large to come from other fingers. He could tell right or left by the slope of the ridges. Ridge slope on a right thumb impression is more steep on its right side, and vice versa for a left thumb.

Collins's next job was to search through his files, paying special attention to the prints of housebreakers who had an arch on the right thumb. He fingered his cabinet's cards slowly and meticulously, for he knew that public acceptance of fingerprint evidence could be won through their successful use in this case. But no luck. On Tuesday morning, the day after the murder, he reluctantly reported to Macnaghten that the print on the cash-box tray did not match any prints on file.

The news was not all bad, however. Collins had compared the cash-box tray print to those of Mr. and Mrs. Farrow and of Sergeant Atkinson, who had mistakenly touched the tray. The print belonged to none of them. That meant that it probably belonged to one of the murderers. If so, Macnaghten and Collins thought they could use it to win both their case and their much desired public respect for fingerprints. But first a suspect had to be found.

. . .

The investigation's first lucky break came when Chief Inspector Fox encountered Henry Jennings, a milkman, and Edward Russell, his eleven-year-old helper. During their rounds, about 7:15 on the morning of the murder, Jennings and Russell saw two men coming out of Chapman's. One had a dark mustache and wore a blue suit, black boots, and a bowler hat. The other was clad in a dark brown suit, gray cap, and brown boots. Jennings shouted to them, "You have left the door open." The mustached man turned around and said, "Oh! It is all right; it don't matter," and left the door ajar.

Fox now had descriptions of two suspects. But if the milkmen had last seen the door open at 7:15, and William Jones arrived at 8:30 to find it locked tight, who closed the door? Was there, Chief Inspector Fox wondered, a third robber who came out after the other two, closing the door behind him?

The fact that three masks had been found in the shop seemed to confirm this theory. Also, three men, two of them answering the descriptions given by the milkmen made a twenty-minute visit to Deptford's Duke of Cambridge Pub at 6:00 A.M. on the morning of the murder. Could the third man in the pub have been the door-closer? The police took his description from the bartender and began searching for the third man, too.

Then police found another witness, Alfred Purfield, a painter. On the morning of the murder, he had waited for a colleague across the street from Chapman's and watched the door being shut. It was "an old gentleman," he told police. "He had blood on his face, shirt and hands. He stayed at the door for a short time and then closed it." The door had been shut by Farrow himself. He had obviously regained consciousness one last time and, too dazed to call for help, simply closed the door before expiring in the parlor. This destroyed the third-man theory. Chief Inspector Fox was not happy. It was three days since the murder and he'd run out of leads.

Enter Fox's second lucky break, a witness who took her time coming forward because she didn't think what she saw was important. Ellen Stanton was on her way to catch the 7:20 train to London on the morning of the murder when she saw two men running from the High Street. What were they wearing? Fox asked. Stanton said one wore a dark suit and a dark cap and the other wore a bowler. Fox's heart skipped a beat. Stanton was wrong about the importance of what she'd seen. Her description matched perfectly with the milkmen's. Did you recognize them? Fox asked. "I recognize [sic] one of the men as Alfred Stratton. . . ." Stanton said. "I don't know the man who was with him. . . ."

Suddenly, Fox had one suspect for sure and guessed he had another. Twenty-two-year-old Alfred Stratton's younger brother Albert, twenty, was his constant cohort and he had a mustache to boot, matching the milkmen's descriptions. The brothers had no criminal records, but they were known by the local police to be living off prostitutes. Fox reported all this to Macnaghten. The Strattons, Fox believed, were the culprits, but he lacked ironclad evidence. Macnaghten nevertheless ordered him to arrest the brothers. Once the Strattons were captured, Macnaghten reasoned, one of their thumbprints would provide all the evidence that was needed.

On Sunday night, six days after the murder, Alfred was arrested at the King of Prussia Pub in Deptford. The next morning, Albert was collared on a Deptford streetcorner. But at the police station, things took a nasty turn for Fox and Macnaghten when neither Jennings, the milkman, nor Russell, his helper, could pick the Stratton brothers out of a crowd of prisoners in the exercise yard. There would be no question, either, of identification by Mrs. Farrow; she had succumbed to her injuries and died. The Strattons, watching the police case fall apart, were so giddy with excitement that they joked that Detective-Inspector Collins tickled them when he took their fingerprints.

With no eyewitnesses linking the Strattons to Chapman's, Macnaghten had to virtually beg the magistrate at Tower Bridge Police Court to remand the brothers into custody. He needed time, he explained, for Collins to compare their prints with the smudge on the cash box. The counsel from the public prosecutor's office warned Macnaghten that the evidence in hand was insufficient for a prosecution. If the prints didn't match, the brothers would go free.

Back in his office, Macnaghten waited impatiently for the results of Collins's examination. Two tense hours passed as he pondered the press-lashing the Yard might take for another unsolved murder. Then Charles Collins rushed through his door, ecstatic. "Good God, sir," he exclaimed, "I have found that the mark on the cash-box tray is in exact correspondence with the print of the right thumb of the elder prisoner."

. . .

The Yard had its murderers. But knowing who committed a murder is a far cry from convicting him for it, especially in a tricky case like this one. No English jury had ever been asked to send men to the gallows on the basis of what was, after all, only a smudge of sweat. Prosecution was a gamble. If the case was lost, the Yard stood to take a considerable public hammering. On the other hand, winning could lead to public acceptance of the greatest crime-fighting tool of its time. Macnaghten deferred to Scotland Yard Commissioner Edward Henry to weigh the odds.

To its credit, fingerprinting had its four-year record of success in identifying habitual offenders. But the Fingerprint Branch had its detractors. Ten fingerprints may identify a man, believed a number of distinguished scientists and doctors, but they highly distrusted the use of a single fingerprint, especially when a hanging was at stake. So strong was their distrust that they would be willing to pit their reputations against Scotland Yard in any upcoming trial.

Most damaging among their mingled grumblings rang the voice of Henry Faulds, the man who first suggested fingerprints to identify criminals. Faulds had compared many thousands of fingerprint sets to satisfy himself that no ten fingerprints could be duplicated on two different people. He complained publicly that no one, including the Yard, had made a similar comparative study to prove that each single fingerprint was unique. Until this was done, he insisted, no man should be sent to the jailer or the hangman on the basis of a single fingerprint, particularly one identified by the Yard's Fingerprint Branch. Ever since the Branch had denied Faulds's part in the fingerprint conception, Faulds had bitterly questioned its integrity. It didn't help the Yard's case that Faulds delivered his arguments with the force of a man who had been scorned.

It didn't help, either, that science didn't have the foothold in the courtrooms that it does today. For most of history, the only evidence allowed at trial was the testimony of eyewitnesses. The use of physical evidence to reconstruct events had been considered too vulnerable to manipulation. The legal process had since been dragged slowly forward, but juries were still more used to hearing what people had seen with their own eyes than what experts said they could deduce by other means. Unlike the rest of society during the industrial revolution, the judiciary had not yet been won over by science. When he decided to take the gamble and prosecute the Stratton brothers for the murders of Thomas and Ann Farrow, Edward Henry knew that this trial could change all that. But the big question remained: Had thirteen hundred years of British legal history prepared the courts for one of their greatest-ever leaps into the future?

Two

To Catch a Crook

In the commons of a sixth century village of what is now France, peasants crowded around a large cooking pot as if expecting a feast. Only boiling water filled the pot, though, and the only thing to feast upon would be a spectacle. Two arguing clergymen intended to plunge their unprotected arms elbow-deep into the scalding water. The idea was to settle a debate they'd had by conjuring supernatural "evidence" of God's judgment. Presumably, the man with God on his side would be protected from the burning temperature.

The argument between the clergymen, a Catholic deacon and an Arian priest, was over the hierarchy of the Holy Trinity. The Catholic said that Christ and God were the same. The Arian insisted that the Son was inferior to the Father. This philosophical difference had caused the separation of their churches two centuries earlier, so in their hotheaded debate over this tired issue, neither clergyman was willing to budge from his original position. Finally, the Catholic deacon challenged the Arian priest to settle the issue in an "ordeal by boiling water."

The priest and the deacon each spent a sleepless night on his knees in prayer, trying to curry God's favor. The next day, they joined the crowd of curious peasants by the side of the cauldron. The flames beneath the pot leaping high, a ring was tossed through the clouding steam and into bubbling water. Each cler-gyman would take a turn fishing out the ring, and victory would go to whomever emerged from the trial least injured.

The Catholic deacon, feigning politeness, gestured his adver-sary forward for the first attempt. The Arian shook his head. The "privilege" of going first, he said, belonged with the chal-lenger. Tentatively, the deacon stepped forward. He looked into the pot, hesitating. The ring whirled around violently. Catching it would be like trying to grab a piece of straw in a tornado. The Catholic slowly rolled up his sleeve, but his arm, the Arian immediately saw, was smeared with oil. Outraged, the Arian accused the Catholic of cheating. He declared the challenge void.

The Catholic deacon resigned from the ordeal, presumably with heartfelt relief. But the Arian was not off the hook. Another Catholic priest had stepped out of the crowd, insistent on taking the disgraced deacon's place. The Arian's back was against the wall. To refuse the new challenge, he would have to concede the issues of faith that had begun the fracas. Reluc-tantly, he examined the priest's arm and, finding nothing to complain about, signaled him to proceed. The priest plunged his arm into the cauldron.

According to legend, the Catholic kept his arm submerged in the vigorously boiling water for two hours as he grasped for the ring. At last, he snatched it, raised it high above his head, and announced to onlookers that the water felt cold at the bottom and comfortably warm at the top. His hand and arm were miraculously uninjured. Emboldened by his adversary's success, the Arian brazenly tossed the ring back into the water and thrust his own arm in after it. Within a moment, his flesh was boiled off the bone to his elbow. God apparently sided with the Catholic.

This kind of "non-rational evidence," as historians would

call it, settled every kind of community squabble in the Germanic tribes that then overran Europe. Even the fate of criminal suspects was at the mercy of the "trial by ordeal" in the Dark Ages. This judicial use of the ordeal was the great-grandfather of modern criminal proceedings. It was the first chapter in the history of the law of evidence, which would lead, more than a millennium later, to the use of fingerprints. But for now the law had no use for earthly clues. God knew who stole or killed, so gathering facts was irrelevant. Conjuring God's judgment was the trial's only goal.

In Saxon England, a frightened suspect often desperately tried to avoid the trial by ordeal by recruiting community members to swear to his good character, in the hope of convincing a judge to let the suspect off the hook. But since fire and brimstone would rain on compurgators who swore falsely, even the slightest scent of doubt in a suspect's innocence meant his friends and acquaintances turned their backs on him. He'd have to admit guilt or submit to the ordeal, turning his fate over to what was, in essence, an elaborate coin toss.

The ordeal required prolonged contact of the accused criminal's bare flesh with either boiling water or, just as commonly, a lump of red-hot iron. The worse the alleged crime, the deeper the judges made him plunge his hand into the boiling water, or the heavier the lump of red-hot iron they made him carry. The singed flesh was then sealed away in bandages. Three days later, judges examined the wounds and divined the evidence of God's judgment. Healing meant innocence, release, and not so much as a muttered apology for the now crippled limb. The stench of infection indicated guilt and execution.

The ordeal was cruel and arbitrary, but it was better than the mass bloodshed that came with its alternative, vigilante justice. Then, a theft might lead to a fight, which ended in a killing, which was in turn avenged by a murder, which then sparked a clan war. The ordeal, at least, had the virtue of resolving conflict, in a procedure agreed by the community, before it devolved into blood feud. One innocent life might be sacrificed, but tens or hundreds were saved.

But ordeals and their ruthlessness long outlived the dangerous Dark Ages clan wars they were designed to prevent. One form of ordeal, the wager of battle, in which the accused and accuser, or their champions, pummeled each other with wooden staffs, remained common in England into the fifteenth century. If a defendant kept up the good fight from sunrise to sunset, he was innocent. Defeat once again fated him to the hanging tree. Even after this judicial jousting faded from practice, it lingered in the law books until it was resurrected for the last time, amazingly, in 1817.

On the morning of May 27 of that year, in Tyburn, near Birmingham, the dead body of a young woman named Mary Ashford was found at the bottom of a pit near her home. Abraham Thornton, a bricklayer and the son of a respectable builder, was arrested and tried for the murder. The jury found him not guilty. An archaic British law, however, allowed the appeal of a not guilty verdict in cases of murder, and Mary Ashford's brother, wracked with grief over the death of his sister, instituted such an appeal. Thornton was again arrested.

If Mary Ashford's brother could cite archaic law to bring this frivolous second trial against Thornton, his lawyers reasoned, then they could also invoke an obsolete statute. In court, they insisted that Abraham Thornton, a large and strong man, be allowed to answer the charges against him in a duel against his less physically robust accuser. The wager of battle, they maintained, had never been expunged from British law. After much quibbling between the lawyers, the judges came down on the side of Thornton. They ruled that if the trial were to continue, the brother would have to fight. Scared for his life, the brother withdrew his appeal, Thornton got his freedom, and both appeal of murder and wager of battle were struck, finally, from the English law books.

. . .

While the administration of justice relied on divining verdicts from God, methods as sophisticated as fingerprinting, and indeed any form of factual evidence, were a long way off. Developments were slow in coming. The old Saxon judicial system remained in use until in 1215 when Pope Innocent III forbade the clergy from participating in ordeals. Walking out on the procedure, the clergy effectively took God with them. And an ordeal without God was like a courtroom with no judge.

So-called investigating juries filled the judicial void. The juries were community recruits—mayors, sheriffs, and tradesmen—who lacked any notion of legal objectivity. Their investigations often amounted to nothing more than knocking on doors to gather local gossip. Suspects were allowed only to listen mutely, unable to say a single word in their own defense, as the juries recounted their hodgepodge of hearsay before a judge, so an indictment, even if it was based on rumor, was a fast track to the gallows. The fact that evidence still was not examined directly in the courtroom tipped the scales of justice heavily toward the prosecution.

Not until 1504 did English legislation call for witnesses to present their own evidence before a judge, the way they would today. An Act of Henry VII, the first to use the word "evidence," urged anyone who witnessed the crossbow shooting of a king's deer to testify openly at court. The Act's promise of a ten-shilling reward blurred the line between a witness's imagination and his memory, but the Act still led the march toward judicial examination of evidence, and a number of other acts calling for eyewitness testimony soon followed.

There were back steps, however, and a new injustice sneaked into the courtroom on the heels of the new evidence: Only the prosecution could call for testimony. Though the defendant could question witnesses rallied against him, he couldn't call his own witnesses or speak in his own behalf. The accused, if he were allowed on the stand, the rationale went, would lie to save his skin. He would then be condemned to hell for breaking the oath of the witness. Refusing the defendant his day in court, therefore, was a supposedly compassionate means of saving his soul.

With no way to answer charges, the accused was left vulnerable to exaggeration and outright lies. As a safeguard, in serious cases like treason or murder, a judge could not consider the yammerings of a prosecution witness unless another witness confirmed them—if only one witness for the prosecution came forward, then the defendant went free. The bad news for the person in the dock was that if the prosecution could find two witnesses telling the same story, his conviction was automatic, regardless of the judge's personal opinion.

Since incredible weight rested on a witness's testimony, the penalties for perjury were steep—if a liar got caught. But the defendant's only protection from the clever perjurer was the oath of the witness. Breaking it condemned the witness's soul to hellfire. A sixteenth-century English legal handbook, *The Country Justice,* advised judges that the way to squeeze the truth out of witnesses was to frighten them with threats of damnation.

But fear of damnation had no power over some witnesses, particularly if, for example, they were religious zealots championing their faith. After Henry VIII separated the English church from Rome, the struggle between the Catholic and Protestant powers often erupted in plots and scandals that ended in the courtroom. Witnesses in this struggle didn't give a second thought to their oath to tell the truth. In their religious fervor, some, such as the Anglican priest Titus Oates, didn't even mind if their outlandish courtroom lies ended with the death of innocents.

By the time he was twenty-five, Oates, a Baptist preacher's son, had been imprisoned for perjury and dismissed from his post as a navy chaplain. In 1677, under the influence of a fanatically anti-Catholic acquaintance named Israel Tonge, he made a false conversion to Catholicism and became a spy against the Roman church. After being expelled from seminaries in both France and Spain, the following year, he rejoined Tonge in London, where the pair used what Oates had learned to concoct a false account of a vast Jesuit conspiracy to overthrow King Charles II.

Oates swore out the fabricated details of the plot before a prominent London magistrate, Sir Edmund Berry Godfrey. The thirty-nine eldest Jesuits, Oates told Godfrey, had secretly met in London in April 1678 to coordinate their plan to assassinate the King and bring to power his Roman Catholic brother, the Duke of York (later King James II). Their plan, according to Oates, included the rising up of Catholics, the general massacre of Protestants, the burning of London, the invasion of Ireland by the French army, and an uprising against the Prince of Orange in Holland.

After the magistrate Godfrey publicized the story, Oates was granted an audience before the King and his council to recount his allegations. They considered his story preposterous. Not long after, Godfrey was found dead with a short sword piercing his heart. Had he, like his father before him, committed suicide, or had he been murdered by Catholics to silence him? History has never solved the mystery, but the investigating coroner decided murder, and Oates's incredible Popish Plot suddenly had a killing to give it substance.

The capital and the nation went mad with hatred and fear.

Justices everywhere searched house by house for papers con-
firming the plot. The jails swelled with papists. Oates was hailed
as the country's savior. In November 1678, he began testifying
in court, coldly pointing a finger of death at the Catholics he
accused of treason. Eventually, the furor died down, Oates's
prevarications were exposed, and he was convicted of perjury.
He was pilloried, flogged, and imprisoned. But by that point,
purely on the strength of his word, thirty-five innocent men had
already gone to the gallows.

Two things missing from the judicial system allowed this
incredible miscarriage of justice. One was the right of defendants to call their own witnesses to contradict the testimony
against them. The other was what is now called physical or
objective evidence—physical objects related to a case—that
today often serves to confirm or contradict witness testimony. If
they had been known, one type of physical evidence, fingerprints, could have been taken from the hilt of Godrey's short
sword. This might have put an end to Oates's lies. But the
importance of any kind of physical evidence would not be fully
recognized until the appearance of full-time professional police
detectives.

.   .   .

When the world's first official detective force finally opened its
doors in Paris in 1812, only a criminal could get a job there. It
took a crook to catch a crook, believed François-Eugéne
Vidocq, the vivacious founder of the *Brigade de la Sûreté* (Security Brigade), and he had the experience to prove it. A former
outlaw himself, Vidocq rose to chief of the Sûreté because he'd
already helped the police snare countless criminals with his
underworld know-how. The fox could hunt better than the
hounds.

Vidocq's first case followed the theft of an emerald necklace

Eugène Vidocq

given by Napoleon to the Empress Josephine. She discovered the necklace missing, in October 1809, from the small estate outside Paris where she had lived since her estrangement from Napoleon. The Emperor, incensed by the theft, worried that his enemies would accuse him of arranging it. He ordered Police Director Joseph Fouché to find the necklace, even if it meant his whole force combing the back streets of Paris. But Fouché was stumped. The main concern of his 300 undercover police spies had always been sniffing out political enemies of the revolutionary government. They had little experience tracking criminals, and even less idea where to search for the Empress's necklace. Their need for help was Vidocq's door of opportunity.

The son of a baker in the town of Arras, the strong-willed Vidocq, by age fifteen, had already killed his fencing instructor, amazingly, in a sword fight. Their duel was the first in a long string of tussles Vidocq fought over women. Five years later, his jealous rage, after yet another fight, landed him a few weeks in prison. He befriended a peasant there, whose only crime was stealing grain for his starving family, and was moved by pity for him. He helped fake a formal pardon that led to the peasant's release.

The scheme was discovered, and Vidocq's various skirmishes with the law for the first time turned serious. His initial arrest for fighting transformed suddenly into a charge of forgery. At age twenty-two, he faced eight years of forced labor. This time, Vidocq had dug himself a hole he couldn't easily climb out of. Though he quickly escaped from prison by stealing a file, sawing through his leg irons, and slipping away in a sailor's stolen uniform, he now had to live the rest of his life with the mark of an escaped convict. And there were many who would happily turn him in for the price on his head.

Vidocq became a pirate, ransacking English ships, and then traveled France, leading a colorful life as a criminal. Often recaptured and always escaping, Vidocq eventually tired of his renegade life and tried to settle down. Hoping to keep his criminal past a secret from the police, he opened a dry goods store in Paris, but he was often blackmailed by those who knew his true identity. He was in constant danger of being betrayed. He wished for an end to the constant running that began when he forged the poor peasant's pardon. And that was the carrot the police dangled before him in return for the recovery of Josephine's necklace.

Vidocq wound his way through the criminal haunts of Paris, scavenging for information about the necklace. In only three days he discovered the identity of the thief and the location of the jewels. Napoleon, delighted, demanded to meet the strange rogue who found his ex-wife's treasure. In a gesture of gratitude, he ordered that the thirty-four-year-old Vidocq be appointed to a police position worthy of his talents, and the now-transformed Vidocq began his crime-fighting career as an underworld spy. Continuing to pose as a fugitive, he pretended to play an active role in the planning of crimes, but secretly tipped off the police before they were perpetrated. Vidocq's crime-fighting tactics were so successful that, three years later, the police prefect Comte Jean Dubois signed an order establishing the Sûreté with Vidocq at its helm.

Vidocq hired eight assistants, who, in line with his philosophy on criminals catching criminals, were all former convicts with vast underworld knowledge. Their work earned Vidocq rapid acclaim. By 1814, he was made a deputy prefect, and in the year 1817 alone, Vidocq and his expanded force of thirty detectives arrested 812 murderers, thieves, burglars, robbers, and embezzlers.

In his years as chief of the Sûreté, Vidocq singlehandedly launched police procedure out of the Middle Ages and into the nineteenth century. He developed the "undercover" technique, planting in the criminal world agents who kept him one step ahead of his quarry. He instituted an early system of criminal identification, recording the descriptions of each criminal's appearance and method of work. Using plaster casts of crime-scene boot prints, he sent thieves to jail by identifying the tread of their boots. In 1822, before ballistic science began, Vidocq solved the case of a murdered Comtesse with the bullet he removed from her head. He proved that it was too big to have been fired from her husband's gun, but just the right size to have come from her lover's.

Vidocq never hesitated to brag about these exploits, especially while drinking in the watering holes of Paris's most famous writers. Hugo, Balzac, Dumas, and Sue all hungrily feasted on his tales, recounting them in their newspaper columns and novels. Victor Hugo, for example, based both Jean Valjean and Inspector Javert, characters in *Les Misérables*, on the detective. The exposure made Vidocq a celebrity, and his sleuthing methods were studied by police officers around the world. Vidocq's fame gave a kick start to professional police detection, and stories of his use of physical evidence and nascent forensic techniques softened the ground for the eventual introduction of fingerprinting.

Yet the detective force that would introduce fingerprinting, a Sûreté-style branch of London's Metropolitan Police, had not yet been started. Governments throughout Europe envied France's Sûreté, but the British felt that a secret detective force was uncomfortably reminiscent of a police state. Then, in 1842, two London murders caused a public outcry that changed their minds.

One of the murders occurred when a suspect chased by police constables turned and shot two of them, one fatally. That a criminal possessed a gun was virtually unheard of in those times; that he would actually use it against policemen demonstrated a disregard for human life that disgusted even most outlaws. The shooter could only have come from the most depraved of criminal backgrounds. Why was he not known to the police?

It emerged that Thomas Cooper, the shooter, was indeed known to be extremely dangerous, at least at the Scotland Yard, London's police headquarters. He belonged to a violent London gang and had a long criminal record. Local police had no idea that such a dangerous felon was holed up in their neighborhood, however, and they walked right into his loaded gun. This outraged the citizens of London. The Yard might as well have let children swim in shark-infested water. And this was the second example of headline-grabbing police ineptitude in only a month.

One evening a few weeks before, a shoplifter left a tailor's shop followed by two salesmen, staying a few steps behind. They'd seen him surreptitiously slip a pair of trousers under his coat. On the street, they quickly related their tale to a passing police constable, and the three followed the thief to the stables where he worked. They confronted the shoplifter, but he denied the theft, so the constable and the salesmen searched the stables for the trousers. Under the hay, the constable uncovered what at first he thought was a plucked goose. Suddenly, the shoplifter rushed out of the stable, closed and locked the door, and imprisoned his pursuers long enough for him to make his escape.

At first, the constable did not understand why his discovery in the hay had scared the shoplifter away. But when he dug the object from the straw, a terrible realization dawned on him. What he had found was not a goose at all but the headless torso of a woman. Later, a noxious odor in the stable's harness room led investigators to the fireplace, where they found the charred remains of her head and limbs. They also discovered the ax, covered with traces of blood, that had been used to dismember her. The man the constable thought was only a shoplifter had apparently killed a woman and tried to cremate her body, piece by piece. Now he was at large.

The shoplifter's name was Daniel Good. A convicted criminal with a two-year prison record, Good had a reputation for temper and violence, and in a fit of rage, he had once torn the tongue from a horse's mouth. These facts were plainly written in the dusty files of Scotland Yard, yet, again, no one had alerted the local police. The result was that the constable on the scene, with all the dimwitted sluggishness that had lately tainted the reputation of the Metropolitan Police, had been given the slip by a criminal much more dangerous than a petty thief.

The public was furious. Nor did the force redeem itself in the search that followed. More than once, when a tiptoe approach was needed, the clodhopper police alerted Good to their impending approach, sending him back into hiding. Eventually Good was apprehended, tried, and hanged for the murder of his common-law wife, but the Yard was lambasted in the press for its inability to undertake the simplest forms of criminal detection.

After the poor handling of the Cooper and Good cases, the reputation of the Metropolitan Police hit an all-time low. So, on June 20, 1842, the government, under pressure from the police commissioners and spurred by the need to repair a red-faced image, finally gave permission for the experimental establishment of a "Detective Force." It began with twelve policemen, transferred from their normal duties, who taught themselves the work of detectives out of three small rooms in Scotland Yard.

The eventual parent to fingerprinting was finally born. But there would be growing pains. The work of the new detectives was at first unsophisticated. They watched and followed suspicious characters, hoping to collar them in criminal acts. They frequented the haunts of known criminals, sometimes in disguise, drinking and carousing and collecting gossip. They searched and questioned pawnbrokers in hopes of finding stolen goods that would lead them to the thieves.

This was all to the good, but a mature detective force would also have a talent for solving crimes from disparate clues, fitting them together like jigsaw puzzle pieces that, when assembled, revealed a picture of the murderer. Twenty years would pass before British detectives first demonstrated such Vidocq-style sophistication. When they finally did, they received a huge fanfare of press acclaim for their solution of the sensational and difficult case of Britain's first murder on a train.

The victim, seventy-year-old Thomas Briggs, was still alive when he was found between the tracks near the railway bridge at London's Victoria Station on a Saturday night in 1864. He died a few hours later of a fractured skull. Briggs had been riding the train from London to Hackney, where he lived. The empty first-class carriage he had occupied pulled into the station, stained with blood, bearing the marks of a fierce struggle, and containing a hat, a walking stick, and a bag.

Briggs's son informed Detective-Inspector Dick Tanner, who investigated the case, that Briggs's gold watch, chain, and eyeglasses were missing from his personal effects. The bag and the stick found in the carriage belonged to Briggs, the son reported, but the low-crowned black beaver hat did not. Briggs habitually wore tall hats. Tanner presumed the beaver hat to belong to the murderer, and it was his only clue.

Tanner circulated to every jeweler and pawnbroker a description of Briggs's missing jewelry, in the hope that they might lead to the murderer. He also visited the manufacturers of the hat— J. H. Walker of Marylebone—but they did not know to whom they'd sold it. The already meager trail of clues had narrowed to nothing. Then a jeweler named Death contacted the Yard in response to the circular.

Two days after the murder, Mr. Death recalled, a thin, sallow-faced man, a German, had exchanged a gold chain matching the description on the circular for a ring and another chain bearing a different pattern. In a second stroke of luck, a cabman named Mathews, hearing the case details discussed in a pub, remembered that he had seen a jeweler's box bearing the name Death in the room of his former lodger, a German by the name of Franz Muller. Mathews identified the hat found in the carriage as Muller's, and gave Tanner a photograph of the suspect along with the news that he had embarked on a sailing ship headed for New York.

Muller's ship, the sailing vessel *Victoria*, would not reach port for six weeks. Muller had five days' start, but there was ample time to overtake him by steamship. Tanner took the train to Liverpool, embarked, and landed in New York long before the sailing ship was due. On the appointed day, Tanner and a New York City policeman rowed out to the *Victoria* in a small boat as it came into New York harbor. They searched Muller's cabin and found Briggs's watch and hat. Muller was brought back to London and tried.

Only physical evidence—the jewelry and the hat—connected Muller to the dead man. A hundred and fifty years earlier, with no eyewitnesses, a prosecution would have been impossible. But the law had evolved. The judge at Muller's trial explained the use of modern evidence to the jury: "One may describe circumstantial evidence as a network of facts cast around the accused man. . . . It may be strong in parts, but leave great gaps and rents through which the accused is entitled to pass with safety. It may be so close, so stringent, so coherent in its texture, that no efforts on the part of the accused could break it."

In Muller's case, the jury decided that the network of facts was unbreakable, and they sent him to the gallows. The law of evidence had evolved far from the early days of the ordeal. Physical and early forensic evidence now had a role in the courts of law. With detective policing and the law of evidence marching towards the twentieth century, it would be just a matter of decades before police solved cases, like the Farrow murders, using evidence left behind by the ridges that had been on the ends of man's fingers since he first evolved.

⋅   ⋅   ⋅

Thirty thousand years ago, Paleolithic artists painted pictures of their hands over and over on the walls of the prehistoric Gargas cave in southern France. On the dusty rock and clay surfaces, in red and black paint, more than 150 impressions and stenciled outlines of their ancient palms and fingers survive. Among them,

the outline of one artist's hand is repeated again and again.
Missing two fingers, probably due to frostbite, the image con-
jures the feel of his ghostly presence. What did he look like?
How did he spend his days? By making an impression—not a
stylized representation, but a true record of his warm hand
pressing against the cold rock—the stone-age artist left behind,
with the same force as old bones in a grave, a vibrant record of
his existence.

Not only in Ice Age France, but throughout prehistoric
Europe, Africa, Australia, and America, the hand was the subject of some of the world's earliest paintings. To prehistoric people, it symbolized the physical manifestation of the innermost
self. Hungry, and they watched their hands rummage for berries
and roots. Angry, and in their hands they felt the weight of a
fighting club. Through action, their hands gave outward expression to their inner thoughts. Through the sense of touch, they
gave inner experience to outward existence. The hand stood as
gatekeeper between self and other. Its symbolic representation,
the handprint, acquired deep meaning.

Sealing promises with the gods, asserting dominion over ter-
ritory, signaling their maker's existence—these were the probable
functions of prehistoric handprints. Twenty-nine thousand
years later, hand prints still did the same jobs. During the ninth-
century Mayan Empire, the soon-to-be victims of ritual human
sacrifice left bloody handprints on the temple walls to make a
last record of their lives. Ottoman sultans, in the same period,
ratified treaties with handprints made in sheep's blood, a royal
seal signifying intent to keep a promise.

In Europe, the more convenient, less messy alternative to the
handprint—the finger mark—appeared only occasionally, and
not until the last several hundred years. In 1691, 225 citizens living near Londonderry, Northern Ireland, sent two ambassadors

to petition Protestant King William III for compensation for losses they'd suffered during his battle with Catholic James II. The citizens promised to pay their ambassadors, if their negotiations were successful, one-sixth of the amount granted by King William. They signed a covenant to this effect with the marks of their fingers. Though this rare European use of finger marks was reminiscent of the more sophisticated fingerprints that came later, its significance, like the handprint's, was entirely symbolic. The lineations left in the marks by the finger ridges went unnoticed.

In fact, not until the seventeenth century's invention of the first crude microscope, the optic tube, did modern western science make mention of the ridges that run across the gripping surfaces of the hands and feet. One of the first microscopists, Dr. Nehemiah Grew, a physician born in Warwickshire, England, in 1641, whiled his hours away dissecting plants and scrutinizing their magnified innards. A member of both the College of Physicians and the Royal Society by the age of twenty-five, Grew founded the field of plant anatomy and was the first to identify flowers as the sexual organs of plants. He also stumbled upon the ridge detail on the ends of his fingers. He published his findings in 1684, making himself the first scientist known to have observed the fingertip patterns that would later be impressed to make fingerprints.

In 1788, another scientist, J. C. A. Mayers, became the first to observe the facet of finger ridges most essential to their use in identification—their uniqueness. He wrote in his illustrated textbook *Anatomical Copper-plates with Appropriate Explanations* that "the arrangement of skin ridges is never duplicated in two persons." In 1823, a professor at the University of Breslau, Poland, Jan Evangelista Purkyně, in his thesis on the skin, noticed that finger ridge patterns fell into distinct categories, the second most important element of fingerprint identification. The categorization of fingerprints would eventually allow them, once filed away, to be easily referenced again, like dictionary entries classified by letter.

Although Grew, Mayers, and Purkyně anticipated the fundamentals of the fingerprint system of identification, their interest was in the advancement of pure science, not its practical application. They had not realized that their discoveries could be used to identify criminals or as evidence in trials, and fingerprints fell into obscurity for the next fifty years. When they reemerged, it would be thanks to a group of illiterate Chinese workers in a region of India governed by one of the first fingerprint pioneers.

Three

Like Rats
with No Rat-Catcher

In July 1858, William James Herschel was promoted and given charge of a rural subdivision in Bengal, India. At the young age of twenty-five, after five years as someone else's gofer, he was suddenly the final authority on everything from his district's tax collection to its road building. He was mayor, sheriff, and judge all wrapped into one, except he didn't get his position because he was popular, and he hadn't won an election. He had been imposed on the local people by the British Lieutenant-Governor. And the ambitious young Herschel intended to make a splash, a particular challenge because of the period's civil unrest.

At the time, Indian citizens would do anything to make things difficult for the much-hated British administration. They didn't show up for their jobs. They stopped cultivating the British landowners' farms. The didn't pay the rent. Frustrations were great for accomplishment-minded young officers like Herschel. Many of their orders were deliberately disobeyed, and much of the rest had no one to carry them out.

Undeterred, Herschel decided, within weeks of his new appointment, to construct a new road. He negotiated the necessary contracts in the sticky heat at his new headquarters at Jungipoor, up the Hooghly River from Calcutta. One of the deals he struck was with Raj Konai, a contractor, for the supply of road-making materials. Herschel was proud of their arrangement. The terms were favorable to the government. But contractors, no less subversive than the rest of the population, had lately made a habit of breaking their contracts. Herschel worried that Konai might deny his obligations.

Herschel's mind raced as he read over their agreement, penned by Konai in Bengali script. Even this written contract might prove useless, Herschel realized, since contractors had begun to deny their own signatures. Suddenly, it occurred to him "to try an experiment by taking the stamp of his hand . . . to frighten Konai out of all thought of repudiating his signature." This spontaneous printing of Konai's hand would later lead to Herschel's being the first in British history to regularly use fingerprints officially.

Born on January 9, 1833, William James Herschel came from an eminent scientific family. His grandfather William Herschel, an astronomer, discovered the planet Uranus. His father, John Herschel, also an astronomer, invented the sensitized paper on which photographs are printed. As a young man, William James, too, was scientifically inclined, but his father encouraged him to strike out in a new direction, so he joined the Indian Civil Service at the age of twenty. Five years later, his promotion to Assistant Joint Magistrate and Collector came in the wake of the Sepoy Mutiny, a beginning in India's struggle for independence and the reason for the civil unrest in Herschel's new district.

The mutiny began after sepoys, Indian troops employed by the British, protested the recent issue of the new Enfield rifle. To load the Enfield, the ends of its cartridges, believed to be lubricated with pigs' and cows' lard, had to be bitten off. This

William James Herschel

clashed with both Hindu and Muslim dietary prohibitions, and in April 1857, sepoy troopers at Meerut refused to use their new rifles. When they were jailed for their refusal, their incensed comrades rose up and shot their British officers, sparking a murderous rebellion that swept the country.

The British responded with ferocious vengeance. Shipped-in reinforcements took no prisoners, bayoneting to death captured sepoys in frenzied massacres. They hanged whole villages, including women and children, for their perceived sympathy with the mutineers. Even after the revolt was suppressed in mid-

1858, British soldiers lashed sepoys convicted of mutiny to the muzzles of their cannons, and fired cannonballs through their chests. With their bodies blown to pieces, according to Hindu religion, the victims had no hope of entering paradise, making the punishment even more cruel.

The slaughter ended, but the conclusion of what the Indians called "the Devil's Wind" did not halt the population's defiance of the unpopular British ruling class. Terrified of revenge for outright rebellion, they subtly engaged in various forms of civil disobedience, including the breaking of contracts with administrators like Herschel. If the administrators took them to court, the Indians simply repudiated their own handwriting. The British were in no position to insist that a signature written in Bengali had come from any particular hand, especially given the region's volatility.

Hoping to keep his road-building project on track Herschel wanted a signature from Konai that couldn't be so easily denied. "I dabbed his palm and fingers over with homemade oil-ink used for my official seal, and pressed the whole hand on the back of the contract," Herschel wrote in his memoir *The Origin of Fingerprinting*. He made a second impression of his own hand, on a separate paper, and pointed out to Konai the distinctive differences between the two. You may think you can deny your handwriting, Herschel communicated to Konai, but you'll never be able to deny that this outline of a hand and these lines of the palm belong to you and no one else. The scheme worked. Konai delivered the road-making materials as promised.

Herschel, impressed with his newfound ability to frighten someone into honoring a contract, experimented with handprints until he eventually hit upon printing just the fingertips, which was less messy. The Chinese and Japanese, probably the first to make widespread use of fingerprints as signatures, had used them on contracts as early as 600 A.D. Herschel, several

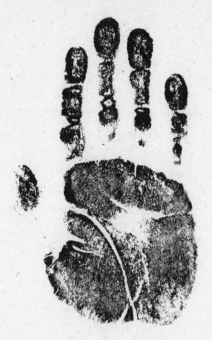

Konai's hand print

authorities have written, likely borrowed the idea from a colony of Chinese living in Calcutta, though Herschel always insisted that the fingerprint conception had come in a sudden flash of his own inspiration.

In 1859, Herschel began collecting, as keepsakes, the fingerprints of his friends, colleagues, and family. Each impression, Herschel noted, was different, convincing him, over time, that fingerprints were unique to each individual. His notebooks included fingerprints taken from the tiny fingers of babies, from Indian nobility, and from old college friends, all dated and labeled by name the way some people collect autographs. He even took the inked impression of a dog's nose: "a little white and black terrier at 2 months." (Much later in history, the inked imprints of the skin patterns on the noses of cattle and horses would also be found to be individual and used to identify them as a safe-guard against theft.)

In 1860, Herschel came up with another application for his fingerprint idea. In Nuddea, near Calcutta, where Herschel took a position as magistrate, the landowners had been turfing the tenant farmers off the land for non-payment of rent. The farm-ers, who cultivated indigo, the primary ingredient of blue dye at the time, couldn't pay because the landowners had not dis-counted rents in line with an indigo market decline. Disputes between tenants and farmers erupted at first into riots, and later into the courtrooms of magistrates like Herschel.

Tenants, desperately clinging to their land, insisted that land-lords tried to collect much higher rents than they'd agreed on in their leases. They presented the supposed documents as evi-dence, but many of them turned out to be forgeries, made par-ticularly hard to detect because they were impressed with replicas of the landlords' seals. Herschel, frustrated by the flow of worthless paper through his courtroom, concluded that land-lords should throw out their seals and instead authenticate leases with fingerprints. He set out to develop his fingerprint sig-nature idea for widespread use.

He concerned himself first with insuring that fingerprints could not be forged like the landlords' seals. He commissioned artists around Calcutta to copy his fingerprint, but none made even a close facsimile. In anticipation of the businessmen's objections to the messy application of ink to their fingers, he wrote in 1862 to his much more practical brother-in-law, Alexander Hardcastle, and asked him to "devise an utterly sim-ple device for inking the finger."

Finally, in 1863, when the non-payment of rent had reached crisis point and land and lease litigation choked the courts, Her-schel penned an official letter to his superiors suggesting his system for prevention of lease forgery. The first two fingers of both the landlord and the tenant should be impressed on each lease, he wrote, so that neither could alter it or disavow it in the future. Government higher-ups rejected Herschel's idea, feeling that it might cause ill feeling just at the time when the indigo disturbances were quieting down. Herschel bided his time.

Fourteen years passed before a more senior Herschel, now magistrate of Hooghly, near Calcutta, was finally able to institute fingerprinting under his own authority. He introduced the system in three separate departments. For a year-long period, between 1877 to 1878, government pensioners in his region signed for their monthly payments with fingerprints. At the registry of deeds, land owners impressed fingerprints to authenticate their transactions. At the courthouse, convicts were forced to fingerprint their jail warrants so hired substitutes could not take their place at the prison. One year before he retired and moved back to England, nearly twenty years after he first came up with the idea, Herschel had finally put fingerprints to official use.

Herschel had, with a little help from the Chinese, conceived the use of fingerprints to irrefutably identify documents with their signatories. But he did not realize until much later, when it was pointed out to him, that fingerprints could be used to identify unknown criminals. Nor had he developed the fingerprint concept sufficiently to be used for that purpose.

Nowhere in his writings, for example, did Herschel mention any large-scale experiments to determine for certain that no two fingerprints were alike. Nor did he discuss what features of two fingerprints should be compared to determine if they had come from the same or different fingers. In fact, the fingerprints in the record books from the Hooghly Registry of Deeds, made in runny, water-based ink, were so faint and smeared that they were often indistinguishable. Even if Herschel understood the technical nuances of fingerprinting, it is clear that his subordinates did not. Under Herschel, fingerprints were more effectively used as a means of intimidation than for any real scientific purpose.

In his 1917 memoir, Herschel would nevertheless claim sole credit for conceiving the fingerprint method of criminal identification, even denying the contributions of the Chinese. As supposed documentary evidence, he produced what was to be known as the "Hooghly Letter," written by him in August 1877 to Bengal's Inspector of Jails and Registrar-General. In it, he suggested the widespread expansion of the two-digit fingerprint signatures he used in Hooghly to jailers' warrants and deed registries throughout Bengal. His suggestion was rejected. More importantly, his letter suggested neither the use of fingerprints to identify unknown criminals in police custody nor their use as crime-scene evidence. Herschel's letter did not suggest the fingerprint system as it is used today.

In 1878, when Herschel returned to England permanently, his successor in Hooghly did not see the value in Herschel's fingerprint registration, and discontinued it. After only one year, Herschel's system fell into disuse. It had not proved itself to anyone but Herschel himself. So, though it was already being quietly investigated by the obscure Scottish medical missionary Henry Faulds in Japan, fingerprinting again fell temporarily into obscurity. This time it did so right when jailers, police, and criminologists needed a system of identification more than ever before.

. . . .

". . . Lawrence Earl Ferrers, Viscount Tamworth, shall be hanged by the neck until he is dead and . . . his body will be dissected and anatomized," said a writ of execution read out in the British House of Lords in May 1760. When Earl Ferrers's wife left him because of his bouts of drunken violence, a man named

Johnson got the job of collecting her maintenance payments. The earl grew to hate Johnson and his monthly visits, and eventually shot him dead. It was for this that Ferrers was tried and condemned by the House of Lords.

At the appointed hour, the noose descended over the earl's head, the gallows trap door swung open under his feet, and he fell until the rope jerked him to a sudden stop. His neck broke with a sickening crack. After his body hung limp and lifeless for the customary hour, undertakers carted it to Surgeon's Hall in the City of London for dissection. Surgeons slit open the abdomen and removed his bowels. They sliced two strips of flesh from his chest and drew them open like curtains to reveal his bloody organs. His eviscerated body, then displayed in a public gallery as a warning against would-be murderers, became a cheap, gory sideshow for the public to parade past. Earl Ferrers's memory suffered its final insult.

The gutting, a fate reserved especially for murderers in eighteenth-century Britain, numbered just one among the many ruthless provisions of the period's criminal law, known as the "Bloody Code." For over 200 different crimes, the Code pre-

The hanging of Tamworth

scribed death as easily as today's law might call for community service. Begging, if you were a soldier or sailor, could earn you a stretch of the neck, and so could spending more than a month with gypsies. Between 1805 and 1818, a fifth of those who mounted the gallows' steps under the Code had done nothing worse than forge bank notes.

Continental society was just as cruel to its criminals. Three years before Earl Ferrers's gutting, France sentenced Robert-François Damiens to be burned and cut to pieces for trying to stab Louis XV. Each time red-hot pincers tore off a piece of Damiens's flesh and opened a new wound, molten lead was poured in to stanch the flow of blood. Letting Damiens bleed to death would be far too kind. "My God, have pity on me. Jesus, help me!" Damiens moaned. When the executioner finally tethered six strong horses to Damiens's arms and legs, his body proved too strong to be pulled apart. Only after his tendons were cut did Damiens's limbs tear from their sockets.

Not all eighteenth-century criminals suffered such endless torture. In lower-profile cases, judges sometimes broke from the law and showed mercy. But this discretionary sentencing turned the judicial process into a sort of high-stakes crap shoot. For the same crime, depending on the judge, one lucky outlaw might be exiled to America, while another might be tortured or killed. This uneven application of the law undermined its moral authority. It was for this reason, not because of compassion for the condemned, that Europe's great legal thinkers finally called for change.

In 1764, the Milanese statesman Cesare Beccaria published *Dei deletti e delle pene* (*Crimes and Punishment*) a seminal book on criminology. It sparked a hundred years' worth of legal reforms, leading, eventually, to a system that could not operate without an infallible method of identification, such as fingerprinting. A twenty-six-year-old aristocrat, trained in law at the

University of Pavia, Beccaria argued that, because of piecemeal development over several centuries, criminal law was an irrational mishmash. Prescribed punishments bore no relation to the seriousness of their crimes. "Whomsoever sees the same death penalty, for instance, decreed for the killing of a pheasant and for the assassination of a man . . . will make no distinction between the crimes," Beccaria wrote.

Criminal law needed a massive overhaul. Beccaria called for standardized punishments that were only severe enough to make would-be criminals think twice. The certainty of a punishment, not its severity, had the greatest deterrent effect, he said. A burglar, positive of being caught and sent to jail, even for a short time, was less likely to commit a robbery than one who, *if* caught, *might* be executed.

Beccaria's writing inspired humanitarian reformers across Europe. In England, philosopher Jeremy Bentham took up Becarria's cause in a 1789 book of his own, *An Introduction to the Principles of Morals and Legislation.* He argued that the object of all legislation should be the "greatest happiness of the greatest number." A punishment should not inflict any more unhappiness than the crime it was designed to deter. By this standard, executing thieves and other petty criminals was immoral.

For one of Bentham's disciples, Samuel Romilly, the end of the death penalty became a quest. The Member of Parliament campaigned tirelessly to reform the Bloody Code and to rid the law of its overbearing cruelties. In 1808, he won a victory when he championed legislation abolishing the death penalty for pickpockets. But Romilly didn't live to see the other fruits of his labors. Heartbroken by the death of his wife, he committed suicide in 1818 at the age of sixty-one.

Between 1832 and 1834, the English Parliament abolished the death penalty for shoplifting a value of five shillings or less, forgery of coins, returning from deportation, letter-stealing, and religious sacrilege. By 1861, only four offenses would be punishable by death: murder, treason, piracy with violence, and arson of royal dockyards. The hangman had seen his day.

Around the continent, prisons sprang up to house criminals spared by the less-often-employed gallows. England's first national penitentiary, Millbank, in London, locked the cell door on its first prisoner in 1816. Pentonville Prison came in 1842, and by 1848, around the country, England had erected fifty-four new prisons, providing 11,000 new cells. In the previous century, prisons had housed only debtors and unfortunates awaiting their turns at trial or the gallows.

Early in the reign of the jailkeeper, in the 1820s and 1830s, crime statistics made their first appearance. They revealed the existence of a breed of hardened outlaws who, no matter how often they went to jail, always returned to their villainous ways. As a social phenomenon, the group quickly attracted the interest of science. Why would this group, in spite of the risks, return

Millbank Prison

again and again to their lawbreaking? Were they bad in their very essence? Or was society somehow to blame?

One of the world's first demographers, the Belgian Lambert Adolphe Quételet, took up these questions. Quételet analyzed three years of French crime statistics, and he published his findings in his 1835 book *Sur L'homme* (known in English as *A Treatise on Man, and the Development of His Faculties*). A third of murders, he found, occurred during barroom brawls. Young working-class men accounted for the greatest proportion of crime. Upper-class villains tended more toward personal violence than theft.

His great criminological discovery was the connection between crime rates and social conditions. When the economy dipped, law-abiding citizens suddenly started stealing. Old thieves stole more often. Crime waves and economic recessions correlated so closely that felons appeared to have no free will. It was as if, in bad times, some societal puppeteer began pulling their strings. Quételet concluded that the blame for lawbreaking belonged partly to society. The severity of a criminal's punishment should therefore depend on the circumstances of his crime.

This penal philosophy begged a question. Was a particular outlaw forced into crime by social conditions? Or did he commit crime by nature? The answer, wrote Arnould Bonneville de Marsangy, depended on the length of the criminal's rap sheet. In 1844, Bonneville published *De la Récidivé*, the first European text to focus exclusively on the habitual offender. Punishment, he wrote, should be tailored to the "perversity of the delinquent." The longer the rap sheet, the greater the perversity, the harsher the punishment.

The ideas of Bonneville and Quételet flew in the face of Beccarian jurisprudence. Beccaria had called for uniform sentencing. For the same crime, each offender should do the same time. Judges were allowed no discretion. But Beccaria's system, despite its attempts at consistency and fairness, introduced new injustices of its own: Treating a starving youth with the same severity as the most callous repeat criminal was hardly an even-handed approach.

In England, in the 1840s, Matthew Davenport Hill, a lawyer and penologist, called for "indefinite" sentencing. Reformed convicts should be released when they demonstrated good behavior over a period. Jailhouse troublemakers who did not reform should remain in prison for life. In 1853, Parliament passed the first Penal Servitude Act, which adopted the "ticket-of-leave" system, a variation on Hill's proposal. This old-world version of parole allowed early release for the well-behaved.

The 1853 Act also brought an end to the deportation of criminals to penal colonies. Sending offenders to Australia had increased in proportion with the decline in executions, until the practice peaked, in the 1830s, at 5,000 convicts a year. But the system was expensive, and Australian colonists did not appreciate their land being filled with convicts. Deportation decreased to a trickle in the early 1840s. After the Penal Servitude Act, England's worst criminals, once sent to Australia, were now sentenced to penal servitude in English prisons. Eventually, they were released and returned to the community.

With the end of both execution and deportation, by the 1860s, thousands of former convicts and ticket-of-leave men lived in London. This frightened the public. Cruel as it was, the gallows did have one merit: convicts, hanged and buried, committed no more crimes. Released convicts, on the other hand, often would. When, in 1860, an old widow was bludgeoned to death in her own home, the public panicked.

This was four years before the Briggs train murder and two years after William Herschel first impressed a handprint in India. Seventy-year-old Mary Emsley owned a number of rental properties in Stepney, London. Each week, she knocked her way from tenant door to tenant door collecting rents, and hauled home a heavy black bag filled with coins. Sometimes her neighbor, a cobbler named Emms, assisted her. She also employed a caretaker, George Mullins, a former constable in the provincial police who was handy with the trowel and the paint brush. Mullins and Emms were Mrs. Emsley's only close acquaintances.

One Monday evening, Mrs. Emsley hobbled home from her rent collecting, and didn't open her door for the next four days. On Friday, her neighbor Emms called the police. They broke through the door and found Mrs. Emsley dead and bloody, her home burgled. Scotland Yard assigned the case to Detective Sergeant Tanner, who four years later would ingeniously solve the Briggs train murder.

A month went by, and Tanner still had nothing to show for his work. Then, one morning, Mullins, the caretaker, told Tanner he had discovered a suspect. As an ex-constable, Mullins explained, he was anxious to assist, so he had pursued inquiries of his own. Over the course of days, Mullins had seen Emms make secret visits to a nearby field. That very morning, Mullins told Tanner, Emms had stealthily retrieved a parcel from the field and brought it to his cottage. Ten minutes later, he emerged with a smaller parcel that he hid in a disused shed nearby. Emms, Mullins concluded, was slowly trying to dispose of Mrs. Emsley's stolen property.

The next day, Tanner visited Emms with Mullins and some other police officers. They searched the shed, but found nothing. Mullins was indignant. "You haven't half searched," he said, pointing. "Look behind that damn slab." The police removed the slab and found a parcel. Unwrapping it, they found some spoons, a check, and other articles of Mrs. Emsley's stolen property. "Are you satisfied?" demanded Mullins of the detective. "Quite," responded Tanner. "I arrest you, George Mullins, on suspicion of having murdered Mary Emsley."

Tanner, it turned out, had already discovered that Mullins was the last to be seen leaving the dead woman's home, but Tanner had no clear proof that Mullins was the murderer. After Mullins's arrest, the police searched his house and found some of Mrs. Emsley's stolen property. The case solved, Mullins was convicted and hanged, but that did not calm the public. The underworld class thrived like rats with no rat-catcher. They overran the capital. Catching them after they committed their crimes was no consolation. After all, what good does it do you if your murderer is caught after you are dead?

The panic over Emsley's murder had hardly died down before a string of violent attacks in the capital started a new scare. In broad daylight, garroters throttled their victims with string, often until they were unconscious, and took their money. Concentrations of police in the affected neighborhoods eventually brought the epidemic to a halt. Many of the arrested garroters turned out to be old convicts released from prison, and this outraged the public. They wanted these ruthless criminals kept off the street.

Responding to public concern, in 1869, Parliament passed the Habitual Criminals Act, providing longer sentences for more hardened criminals. A first-time offender, the rationale went, might merely be a weak character faced with desperate circumstances—the type of criminal whom Quételet and Bonneville said was the victim of societal conditions. To this otherwise decent citizen the judicial system delivered a short, sharp shock, sufficiently deterring him from offending again.

Habitual offenders, on the other hand, were "a criminal class distinct from other civilized and criminal men," wrote the Scottish prison surgeon J. Bruce Thompson in 1870. Like communities of fishermen and miners, Thompson argued, habitual criminals lived and intermarried among themselves. Their inbreeding caused the transmission of the criminal bent to their offspring. An Italian professor at the University of Turin, Cesare

Garroters at work

Lombroso, in 1876, seemed to confirm the concept of hereditary criminality, believing that even their physiques differed
from other men's. He concluded that habitual criminals were
marked by heavy jaws, receding brows, and pointy crowns.

Since the habitual criminal was obviously born, not made,
the thinking went, he had no power to change. Nor could he
have his criminal bent punished out of him. Only quarantine
from the population would prevent his future felonies. For these
reasons, the Habitual Criminals Act gave judges the right both
to hold a crook in jail while police inquired into his previous
history and to lengthen his sentence if he turned out to be a
recidivist. The length of a criminal's rap sheet finally determined
the length of his prison stay. But how would the system distinguish the habitual from the first-timer? Only their court records

distinguished them. Adopting false names, habituals eluded identification with their past misdeeds, pretended to be first-time offenders, and sometimes avoided long-term removal from society. Without some means of reliable identification, the new system would never be effective.

Since the Industrial Revolution swept peasants off the land and into cities like London, criminals, once well known to law enforcers in their townships, now roamed anonymously among the urban throngs. No longer were their identities marked by their families, villages, or trades. Without these community ties, officials had no way to put names to the unknown faces that came to their attention. Identifying people in the newly urbanized society was close to impossible.

In 1874, when a man claimed to be the long-lost heir to the Tichborne baronetcy, the law had no scientific way to confirm or deny his identity. Police had adopted registers of photographs and distinctive marks—tattoos, scars, and moles—but the success of these systems was sporadic. They were no use at all in the case of the Tichborne "Claimant," as he came to be called, which dragged through the courts and choked up the legal system for nearly three years.

Twenty years earlier, in 1854, Roger Tichborne disappeared at the age of twenty-five. He had traveled through South America in self-imposed exile ever since his secret engagement to his first cousin Katherine had been discovered by her father Sir Edward, the then baronet of Tichborne. He had forbidden their marriage. When news came that Sir Edward had died, Roger, now first in line to the baronetcy after his father, rushed home in hopes of wedding his love.

In April 1854, he set sail on the *Bella* from Rio de Janeiro to Jamaica, where he planned to find passage to England. The *Bella* never arrived. Six days after she set sail, a ship crossing her intended course was barraged by flotsam, including an over-

Roger Tichborne

turned lifeboat bearing the ill-fated vessel's name. No survivors were found. In a squall, the *Bella*'s cargo was presumed to have shifted, capsizing the ship, sinking it, and taking down everyone on board.

For two years, the Tichborne family hoped that survivors had been picked up by some unknown ship, but with time they faced the truth—all except Lady Tichborne, Roger's mother. Years later, she still refused to give up hope. Unscrupulous sailors, to earn a few coins, regaled her with false tales of rescued *Bella* passengers. When Sir James, Roger's father, died in 1862, there was no one to chase these liars away, and Lady Tichborne was ripe for deception.

In November 1865, she received a reply to one of the newspaper advertisements she had placed around the world in search of her shipwrecked son. The correspondent, Arthur Orton, earned his living as a butcher, and his wife was an illiterate servant girl. Pretending to be Lady Tichborne's son, he wrote to her from Australia, where, he said, the ship that rescued him

Arthur Orton—"The Claimant"

had been destined. When Orton found an old Tichborne servant living in Australia who, for a fee, professed recognition of his dear young master, Orton's deception was complete. Lady Tichborne beckoned him to Paris, where she was living, and Orton went, but not before the old servant had schooled him in the history and traditions of the Tichborne family.

Orton first met Lady Tichborne on a dark January afternoon in his hotel bedroom in Paris. He had caught a chill, couldn't get out of bed, and was too confused by his fever to keep up his deception properly. He spoke of a grandfather whom the real Roger had never met, of serving in the ranks of the army, though Roger had been an officer, and of going to school at Winchester when Roger attended Stonyhurst. Orton was also obese. Twenty years earlier, Roger had been rail-thin.

The deluded Lady Tichborne nevertheless claimed to recognize her son. Though none of the rest of the Tichborne family was fooled, the mother never changed her mind, even after living with Orton and his wife and children for weeks. She allowed her "son" an allowance of £1,000 a year, but Orton was not satisfied. Four years after his "return" to England, in 1870, he began a suit of ejectment against Sir Alfred, the infant who inherited the family baronetcy when Roger was lost at sea. In a 102-day trial, Orton produced over one hundred witnesses who swore to his being Roger Tichborne. Only seventeen witnesses testified against him. His case seemed won. Then began his own twenty-two-day testimony.

The Claimant could answer no questions on the first sixteen years of Roger's life. He could say nothing of Stonyhurst. He offered no recollections of Roger's books, music, or games. He claimed to have forgotten his mother's maiden name. He couldn't, like Roger, speak or write French. Nor did he remember anything about the secret engagement to Katherine, the most important thing in Roger's life. The Claimant's suit was rejected by the jury. The judge immediately began proceedings against him for perjury.

The criminal trial against the Claimant began in April 1873. Orton's chief defense was a witness who claimed to have been a crew member on the *Osprey,* a ship that he said picked up a lifeboat full of *Bella* survivors on its way to Australia. The Claimant, the seaman said, was among those whose lives had been saved. Shipping records, however, contained no trace of the *Osprey*'s existence, and the supposed seaman, it turned out, had only recently been released from prison. In February 1874 after 188 days of testimony, the jury deliberated for only half an hour before returning a verdict: Guilty. The judge sentenced Orton to fourteen years' penal servitude.

Only with huge judicial and financial resources was the Claimant's imposture finally proved. What would stop future persons from misrepresenting their identities to equally damaging effect? In these modern times, when cities swarmed with unknown faces, cases of misrepresented and mistaken identity threatened to overwhelm the legal system. A little-known Scottish doctor, Henry Faulds, took the problem to heart. In 1873, Faulds was in London, preparing to sail for Japan. He personally witnessed the huge crowds waiting outside the Old Bailey to hear the latest news in the Tichborne case. "From that time," he later wrote, "the question of identification as a pressing scientific problem in medical jurisprudence was never long absent from my mind." Fifteen years later, Faulds would be the first to suggest the use of fingerprints to Scotland Yard.

Four

Marks on a Cocktail Glass

Henry Faulds was an impressionable sixteen years old in 1859 when Charles Darwin scandalized Victorian society with his book *On the Origin of Species by Means of Natural Selection.* According to religious thinking of the period, the earth's plant and animal species had been created, fully formed, by God, either when He made the earth or, later, to take the place of other species that had gone extinct. But Darwin, after exploring along the coasts of South America aboard the HMS *Beagle,* had come to a different conclusion.

On the Galapagos Islands, he discovered varieties of finches, mockingbirds, and tortoises that existed nowhere else in the world. In continental South America, he found fossils of armadillos that slightly differed from living specimens. His findings demonstrated that, somehow, species varied in form from place to place and from era to era. They were not, as the evangelists theorized, fixed in a Creationist mold. Species transformed.

The manner of the transformation, according to Darwin's theory of evolution, was the process of natural selection—survival of the fittest (though this phrase was never used by Darwin himself). Within any species, Darwin noticed, the physical characteristics and abilities varied slightly from individual to individual. He asserted that a species' members with characteristics least suited to the environment—because they couldn't as effectively compete for food, escape predators, or contend with disasters—were least likely to survive. "Natural Selection," as Darwin called the process, ensured, therefore, that only those best suited to the environment survived and mated, transmitting their characteristics to their offspring. Over hundreds of thousands of years, Darwin concluded, this process caused identifiable changes in species: evolution. And the implication of Darwin's theory was that man, too, had resulted from such a process of evolution by natural selection.

Darwin's theory threw the Victorian clergy into a rage. It completely refuted their literal interpretation of Genesis. Creation, the theory seemed to suggest, was an accident that occurred, at best, while God stood idly by. The more outspoken clerics ranted about the theory's anti-religious bent at every opportunity. During a highly public debate, one of them, the evangelical bishop Samuel Wilberforce, sarcastically asked the agnostic biologist Thomas Huxley whether "you are descended from an ape through your father or your mother?" Meanwhile, 11,000 British clergy signed a declaration recording their belief "without reservation or qualification" in the divine authority of the Scriptures. Around the country, foot soldiers of science and religion had locked themselves in philosophical combat.

The theologians' disdain for Darwin's theory hurt him deeply. The tension surrounding the conflict caused his chronic vomiting, insomnia, and palpitations. For his part, Darwin did not believe that evolution unfolded in a godless universe, but neither did he believe that God intervened in all affairs. "There seems to me too much misery in the world," he wrote. He postulated, instead, that God created the laws of nature, including evolution, and then left the details to the workings of chance. Ultimately, the whole issue was too profound for the human intellect to grasp. "A dog might as well speculate on the mind of

Newton," he wrote in a letter to the American botanist Asa Gray, who had provided some data for his theory.

Gray, a devout Christian who accepted Darwin's ideas, had urged him to concede to the church at least that evolution was God's tool, wielded for His own purpose. Gray's intellect applauded the theory's scientific truth, but his soul mourned the apparent meaninglessness of its workaday mechanics. Where did God fit in? Gray did not struggle alone with this question. It troubled many of science's more religious advocates, including, eventually, Henry Faulds. The question entered Faulds's thinking in early adulthood, dramatically affecting the rest of his life. Strangely, it led to his study of fingerprints.

Faulds was born on the first of June, 1843, in Beith, a little town in southwest Scotland. His father, William Pollock Faulds, owned a successful business that hauled the harvest of local farmers to the markets of Glasgow. With his profits, he raised his four children, educated his first son Henry at the local private school, and bought stock in the Western Bank. He dedicated his spare time to his duties as an elder in the United Presbyterian Church, and encouraged his children to keep their noses in the Bible.

William's family lived prosperously and happily until 1857, when an economic storm over Britain blew in a steep rise in interest rates. They bankrupted the Western Bank's largest borrowers, forcing the bank itself to close its doors, never to reopen again. The bank's creditors called in its loans, including those to William Faulds's business. He could not raise even a penny on the pound for his shares in the bank. They were worthless, and William was ruined. The Faulds' family wealth was washed away.

The senior Faulds struggled to support his family as a working man, and his thirteen-year-old son, Henry, pitched in. He quit school, and went to work as an office boy in a cotton, tea, and coffee business belonging to Thomas Corbett, Henry's uncle. One year later, while the Faulds family struggled to make ends meet, William Herschel, in India, who was ten years older than Faulds, took the print of Konai's hand. The following year, Darwin published *Origin,* and the year after that, the murder of the old landlady Mary Emsley threw London into a panic.

Around 1860, Tom Corbett moved his tea and coffee business to London. Henry Faulds stayed in Glasgow and took clerical work at R. T. & J. Rowat, Shawl and Dress Manufacturers. Faulds worked there for five years, all the while trying to scratch his way up in the world. He attended private classes at night, until, in his twentieth year, his family had finally recovered financially. His father scraped together his pennies, and Henry enlisted in the Faculty of Arts at the University of Glasgow. In 1868, when he graduated, he enrolled at Anderson's College, Glasgow, and began his study of medicine.

Faulds earned his physician's licentiate, traveled to London to study surgery at St. Thomas's Hospital, then returned to the Glasgow Royal Infirmary to practice under the great medical pioneer Lord (then Mr.) Joseph Lister, eponym of Listerine. As head of the Royal's new surgical block, Lister discovered the "antiseptic method," one of surgical medicine's most important developments. Simply by sterilizing his operating room and instruments, Lister reduced death among amputees from nearly 50 percent to 15 percent. He had similar results with other operations.

Lister's new procedures, Darwin's new theories, and a host of other mid–nineteenth century developments made the era of Faulds's scientific education an exciting one. They also threw Faulds, like Asa Gray before him, into paroxysms of self-questioning. Henry's father, whom Henry loved and respected, raised him to believe in a literal interpretation of the Bible and to dedicate himself to the church. Throughout his studies, Henry taught Sunday school at the Barony Church in Glasgow. Yet the rift between science and religion troubled Henry. Sci-

Henry Faulds

ence, Lister's surgical techniques had shown him, could alleviate man's suffering. If God did not side with science, how could it save so many lives?

In addition to his full-time work as a doctor, Faulds would write two books on travel in the Far East, three others on fingerprinting, and many academic articles, and found three magazines. Two of the magazines, published with funds taken from Faulds's own meager living, would concentrate on the philosophical rift that so troubled him. Faulds wrote, "No man can study modern science without a change coming over his view of truth . . . Science cannot overthrow Faith; but it shakes it. Its own doctrines, grounded in Nature, are so certain, that the truths of Religion . . . are felt to be strangely insecure."

In 1871, Faulds sailed to Darjeeling, India to work as a medical missionary with the Church of Scotland. Whatever his philosophical doubts, Faulds wished to dedicate his life to a good cause. For two years, he treated and cared for the poor in Darjeeling. Then his theological questioning got him in trouble.

Faulds had peppered his talks on western thinking to the local population with references to Darwin and science. This angered the evangelical priest who headed the mission. Back in Scotland, the Church Foreign Mission Committee wrote to Faulds, admonishing him that "they expect all their agents whether ordained ministers or not to teach the doctrines of Christianity in accordance with standards of the church." But throughout his life, Faulds demonstrated a certain hardheaded-ness when asked to compromise his principles. He would not temper his views for the church. Instead, he resigned.

Back in Scotland, in 1873, he applied to a different church, the United Presbyterian, for another missionary position. The Reverend Thomas Dobbie of Leith wrote in a letter of reference that Faulds was able, intelligent, tasteful, devoted as a Christ-ian, and ". . . has never failed to endear himself to those with whom he has come in contact." A letter from the Church of Scot-land said that it "would willingly have retained Faulds' services," and settled any question regarding his conduct in India. On July 29, 1873, Faulds received his letter of appointment from the Pres-byterian Church. In September, he married a girl from Beith, Isabella Wilson. In December, the newlyweds sailed from London to establish the first Scottish Medical Mission in Japan.

Though Roman Catholics had banged their drum in Japan since the sixteenth century, Faulds and his colleagues were among the first groups to fly the flag there for Protestantism. They set up shop near Tokyo, in Tsukiji, an area set aside for foreign inhabitants. At first Faulds ran his medical clinic from a little wooden building, originally erected as barracks for French soldiers. In May 1875 the area around the barracks became infested with frogs. Faulds could not move from one patient to another without stepping on them. He purchased a building more suitable as a hospital, and worked from there until he left Japan ten years later.

Faulds ran his hospital, lectured Japanese medical students, taught Lister's antiseptic methods to Japanese surgeons, trekked into the mountains to heal the bedridden, established a society for the blind, and set up lifeguard stations to prevent drowning in nearby canals. He halted a rabies epidemic that killed small children who played with infected mice, and he helped stop the spread of cholera into Japan. He even cured a plague infecting the local fishmonger's stock of carp. By 1882, his hospital treated 15,000 patients annually. Faulds was inundated.

Yet the Foreign Mission Board of the United Presbyterian Church wasn't satisfied. "You need to remember you are an emissary from Christ," they wrote. They constantly nagged the deluged doctor to pursue what they considered to be "the one aim of his life, that is . . . 'winning souls' to the Savior." In response, Faulds posted the Ten Commandments and the Apostles' Creed on the wall of his waiting room. The doctor chatted about religious matters with his patients and gave them biblical literature.

The more elite Japanese, who didn't visit the mission hospital, were outside Faulds's immediate influence. They thought Christianity was a superstition, and that the westerners who believed in it were unsophisticated. To dissipate their prejudice, Faulds established a magazine he called *Chrysanthemum*, intended to demonstrate that westerners indeed had literary and scientific culture. Meanwhile, an American archaeologist, Edward S. Morse, had begun delivering lectures on Darwinism. Huge Japanese audiences attended. Western science intrigued them more than Western religion.

During his talks, Morse drew atheistic conclusions from Darwin's theories and questioned the existence of God, enraging the missionary community. The clerics needed an eloquent adversary to mount the stage and repudiate Morse's arguments. Faulds, by this time, had resolved his inner conflict between evolution and creationism. He interpreted Genesis symbolically and accepted evolution as the process by which God brought His beings into existence. Though his views weren't to the taste of most clerics, they made Faulds the perfect adversary for Morse. In a series of debates staged in 1878, Morse argued that evolution made God redundant, while Faulds said the process was God's tool. Several thousand Japanese turned out to hear the two westerners fight it out.

Despite their staged antagonism, Faulds and Morse struck up a friendship. Previously a zoologist and professor at Maine's Bowdoin College, Morse eventually taught at Tokyo University and helped establish the Japanese Imperial Museum. He ventured to Japan to study brachiopods, a type of shellfish common in its coastal waters. But within days of his arrival, he was sidetracked by his discovery of an ancient mound of discarded shells and bones, where long-dead villagers had piled their refuse. The shell mounds were the last remnants of "savage races," as Morse called them, who visited the shore to feast on mollusks and fishes.

At the time, academics searched frantically for empirical evidence of man's evolution. Shell mounds helped provide that evidence. In Florida, researchers digging such ancient garbage dumps discovered charred pieces of human bone that appeared to have been cooked—proof, they concluded, of cannibalism. More commonly, shell mounds yielded gouges and needles made from bone and hammers, axes and arrows made from stone. Morse's Japanese excavations were distinguished by cooking pots and other vessels made from clay.

Eighty-nine meters long and four meters deep, Morse's shell mounds lay alongside the Imperial Railway Line, just before

Omari station, about six miles outside Tokyo and half a mile from the shoreline. The water level had dropped dramatically since the ancients had gone to Omari to rummage for seafood. The bottom layers of the shell mound, Morse estimated from the receded waterline, were at least 2,000 years old. Morse's workers carefully sorted through the mounds and carried away basketsful of artifacts. Henry Faulds regularly rode the train to Omari to sift through the booty. He was entranced.

One day, while turning over ancient pottery fragments in his hands, Faulds noticed minute patterns of parallel lines impressed in the clay. He examined them closely, trying to discern their source. Some months earlier, Faulds had lectured his medical students on each of the five senses. During preparation for the lecture on touch, he had noticed the swirling ridges on his own fingertips. In a flash, he realized that the 2,000-year-old impressions he now examined in clay came from the ridges on the fingers of ancient potters.

Did modern potters leave such marks, too? Faulds scoured the contemporary markets of Tokyo, closely examining the sur-

Workers at the Omari shell mounds

faces of current-day pottery. The marks were everywhere. On China tea sets in one market stall he noticed how "one peculiar pattern of lineations would reappear with great persistency, as if the same artist had left her sign-mark on her work." Suddenly it occurred to him that a piece of pottery could be matched to a particular potter by the ridge markings left in the clay. He had begun to suspect that finger-ridge patterns were unique to each individual, the basis for their use in identification. At first, Faulds paid little attention to this detail.

At that time, Faulds did not fancy himself as a detective wanting to identify criminals, but as an anthropologist wishing to throw light on the origins of humanity. Since the 1860s, anthropologists had sought to classify populations according to their physical attributes. Among them, Paul Broca, who founded the Anthropological Society of Paris in 1859, had used measurements of the bony portions of the head and face to distinguish one group from another. By careful analysis, Broca showed, for example, that northern Europeans were distinctively more long-headed than central Europeans. Faulds hoped populations might be similarly classified by finger-ridge patterns. He thought the patterns might differ by race, era, and geography, much like Broca's facial characteristics.

The Scottish doctor studied the fingerprints of his friends, his family, his grocers, even the workmen who came to his house. At first, Faulds examined their finger ridges directly, making sketches for his records. Next, he began recording their fingertips in wax. Finally, he hit on the technique of inking the fingertips and recording their impressions on paper. Twenty years earlier, William Herschel, unknown to Faulds, had begun collecting the prints of the thumb and first two fingers of his acquaintances. Now, Faulds began a similar practice, except for one crucial difference—he insisted on inking and printing all ten of his subjects' fingers, a move that would one day make finger-print sets easier to differentiate in large criminal registers.

Faulds's collection of prints swelled to the thousands, but they all came from European and Japanese fingers. He needed a greater variety to determine whether finger-ridge patterns differed from race to race and area to area as he had postulated. In an effort to expand his data, he wrote more than a hundred letters to scientists around the world, asking their assistance in collecting fingerprints and including copies of specially created ten-digit fingerprint forms. Faulds received almost no response. "Some thought I was an advocate of palmistry . . . most took no notice whatever." Faulds's fingerprint studies had come to a dead end.

Coincidentally, during this period, the supply of medical alcohol at Faulds's hospital, kept in a bottle in a locked cabinet, ran inexplicably low. It had to be restocked again and again before Faulds finally realized that the bottle was emptying itself into some thirsty person's gullet. When he found a makeshift cocktail glass in the form of a laboratory measuring beaker, he examined its surface and discovered a nearly complete set of sweaty finger marks. Faulds searched his collection of finger-print cards for a match, and found one. It belonged to one of his medical students—culprit discovered.

At first, Faulds did not recognize the new use for fingerprints he had unwittingly stumbled upon. Then, a month later, someone attempted to burgle the hospital by climbing up a wall and through a window. Local police accused a favorite member of Faulds's staff, but the ridge patterns in a sooty handprint found on the wall, Faulds found, did not match those of the accused. He showed his evidence to the police and exonerated the staff member.

This time Faulds saw the light. He remembered the crowds

he had seen outside the Old Bailey, waiting for news of the trial of the Claimant. A filed set of the shipwrecked Roger Tichborne's fingerprints, Faulds realized, would have destroyed the Claimant's case in a moment. Similarly, a fingerprint register of habitual criminals would foil their attempts to use false names and get lighter sentences. Faulds's conception was similar, in a way, to that of William Herschel, who, unknown to Faulds, had one year earlier introduced fingerprints' official use in Hooghly, India. Herschel, however, used fingerprints only as a form of signature to authenticate documents. Faulds's idea had much farther-reaching ramifications. He realized fingerprints could solve the problem of identification that so troubled the British legal system.

Faulds was loath at first to publish his idea. He was plagued by a "most depressing sense of moral responsibility and danger. What if someone were wrongly identified and made to suffer innocently through a defective method? It seemed to me that a great deal had to be done before publicly proposing the adoption of such a scheme." Faulds first set out to prove conclusively that fingerprints were unique to each individual and, second, that they stayed the same throughout a person's life.

In one experiment, Faulds and his medical students shaved off their finger ridges with razors until no pattern could be traced. The ridges grew back, without exception, in exactly the same patterns. They repeated the experiment, removing the ridges by any number of methods—by "pumice-stone, sand paper, emery dust, various acids, caustics and even spanish fly"—and each time the results were the same.

Next, Faulds studied infants to see if growth affected their fingertip patterns the way it dramatically changed the rest of their bodies. It didn't. Over a period of two years, he also examined the hands of large numbers of Japanese children and some thirty-five European children between the ages of five and ten. In no case did the ridge patterns vary. When an epidemic of scarlet fever swept through Japan, causing severe peeling of the skin, Faulds again studied the fingerprints and found no before-and-after change.

"Enough had been observed," Faulds decided, "to enable me confidently, as a practical biologist, to assert the invariableness, for practical identification purposes, of the patterns formed by the lineations of human finger-tips." Fingerprints were permanent. Meanwhile, the many thousands of fingerprint sets collected and mutually compared by Faulds satisfied him that each person's fingerprint set was truly unique. He was finally ready to go public.

Faulds's first concern was still to spread the study of fingerprints among ethnologists and anthropologists around the world. To this end, he hoped to enlist the aid of his hero Charles Darwin. "I am an ardent student of your writings," he wrote in a letter dated February 1880. "I trust I may venture to address you on a subject of interest. I allude to the rugae and furrows on the palmar surface of the hand." He explained the purpose of his comparative study of fingerprints, but complained that he was short of samples from around the world. He hoped a word or two from Darwin might set researchers working everywhere.

Darwin, by 1880, was too old to help. He wrote a letter of apology to Faulds, and promised to send Faulds's letter to his cousin, Francis Galton, an esteemed scientist who was interested in using Darwin's theories to improve the human race. In April 1880, true to his word, Darwin wrote: "My dear Galton, The enclosed letter may perhaps interest you, as it relates to a queer subject. You will perhaps say hang his impudence. I have written to Faulds telling him I could give no help, but have forwarded the letter to you on the chance of its interesting you."

Galton replied, saying that "I myself got several thumb impressions a couple of years ago but failed, perhaps from want of sufficiently minute observation, to make any large number of differences. I will do what I can to help Faulds in getting the sort of facts in having an extract from his letter printed." Galton did not keep his word; Faulds never heard from him.

The doctor was not deterred. Eight months after writing to Darwin, he published his fingerprint ideas and pleas for further investigation in the pages of the prestigious scientific journal *Nature*. In his letter in the October 28, 1880, edition, Faulds suggested the use of "bloody finger-marks or impressions on clay, glass, etc." for the "scientific identification of criminals." He also suggested that registers be kept of "the for-ever-unchangeable finger-furrows of important criminals."

Faulds's letter was the first in the scientific literature to suggest the basic concepts of the fingerprint system of identification as we know it today. Much to Faulds's disappointment, it did not spark great scientific discourse. Scientists did not, as Faulds had hoped, fill his mailbox with samples from around the world. Police chiefs did not race to institute his ideas in their departments. In fact, the only notable response was a reply from William Herschel, published about a month later on November 25, 1880, also in *Nature*. Now back in England, Herschel reported his limited use of fingerprints two years earlier in India as a method of signature. Herschel's account of his bureaucratic application of fingerprints did little to further kindle any widespread interest in Faulds's ideas.

In his frustration, Faulds took it upon himself to write to the chiefs of the major police forces around the world. Patiently, he dispatched letters to New York, London, and Paris, among others. His campaign was a lonely one, rarely instigating even the courtesy of a reply. To make matters worse, a second system of scientific criminal identification had been developed by a young clerk named Alphonse Bertillon in Paris. Without knowing it, Faulds ran in a race with Bertillon to see who could be the first to convince a police chief to experiment with his system. Whoever won would from then on be considered the father of scientific identification.

Five

In a Criminal's Bones

Twenty-five-year-old Alphonse Bertillon, Faulds's unknown rival, crouched each day over a desk in a corner of the vast storage halls in the basement of the Paris Prefecture of Police. In the summer, Bertillon's sweat dripped onto his papers in the unbearable heat. In the winter, his feet froze, and his gloved fingers ached as they gripped his quill pen. For the then equivalent of about $250 a year, squinting by the light of a flickering gas lamp, Bertillon scribbled police descriptions into the five million criminal files gathering dust in the massive archive. "Stature: average," they typically said. "Face: ordinary." These files would never identify anyone.

Bertillon fought back the burning frustration and bitterness caused by his work's futility. A sufferer of migraines and nosebleeds, sarcastic and volatile by nature, his bad temper had already cost him jobs as a bank clerk and a tutor. This time, he struggled to maintain his everyday reserve. He kept his movements characteristically slow, and his voice expressionless. He would not display his temper. Bertillon had taken a new, secret lover, and he wanted to prove to her that he could make something of himself.

Born in Paris on April 24, 1853, Alphonse was the second

son in a family of distinguished anthropologists. His father, Louis-Adolphe, was a founding vice president of the Society and School of Anthropology. His maternal grandfather was Achille Guillard, an even more distinguished anthropologist. In 1854, Guillard published one of the first European texts on demography. The two great men worked together and studied, among other things, the physical characteristics of the French population. As a boy, Alphonse naturally learned their "anthropometric" techniques of bodily measurement that would much later in his life become enormously useful. For now, there was little about Bertillon that could be considered successful.

By age six, he'd been kicked out of his first school, Chaptal College, for lack of discipline. When the Bertillons hired a tutor to educate Alphonse at home, the rascal stole the teacher's glasses, hid in a cupboard at lesson time, and generally harassed him. The tutor quit. Bertillon next attended the Rossat Institution in Charleville, a school for difficult chidren. At age eleven, he was expelled from there, too. In his teens, then at the Imperial Lycée in Versailles, Bertillon accidentally lit a fire in his desk. He'd been using a spirit lamp to make hot chocolate for his chums. He clamped down the desk's lid, refused to let the schoolmaster open it, and hit him over the head with a Greek dictionary. Expulsion number three.

Bertillon's family had nowhere left to send him. His father and grandfather had to take the reluctant young pupil under their own wings, where Alphonse had always wanted to be. With them, the boy eagerly pored over demographic texts. He excelled at botany and natural history, and with hard work, conquered Greek and Latin, too. At age twenty, Bertillon passed the baccalaureate in science and literature. His burst of academic accomplishment ended there. Much to his father's chagrin, Alphonse had no taste for further education.

Alphonse Bertillon

Bertillon reluctantly took a job at a bank, but his terrible handwriting got him fired within a month. Disappointed, his father packed him off to England, where at least he would learn English, and cut off his allowance. For the first time, Bertillon stood on his own two feet. He managed to hold onto his jobs teaching French in a boys' school and tutoring in a private family. Drafted into the army, he returned to France, where he also took a medical course at the University of Clermont-Ferrand. He passed the "first medical" with distinction, but soon fell back into his dilatory ways. Then, his life completely changed when, for the first time, he fell in love.

In 1879, probably through his father's elite academic connections, Bertillon met and became besotted with a Swedish noblewoman. The differences in their social stations forever prevented their marriage, but they rendezvoused secretly and wrote each other passionate love letters. Bertillon never revealed the lady's name to his family or friends, but all his life he remained obsessed with her, always keeping her photograph near him. Propelled by a desperate need to impress her, Bertillon finally decided to make a success of his life. He asked his father one more time to help him find a job. In March, he took his place in the dungeonlike halls of the identification bureau of the Paris Prefecture of Police.

The paper treadmill Bertillon stepped into was kept turning not by its usefulness but by the momentum of bureaucracy's large spinning wheels. Even if a criminal gave his true name, finding his file among the alphabetized records was nearly impossible. A search for any one name, Martin Pierres, say, might yield 300 files. How was a police officer to know which file corresponded to the Martin Pierres in custody? If a criminal gave a false name, the situation was even worse. Photographs had been added to the files in the 1840s, but finding one photograph among the 80,000 on file was close to impossible.

In practice, posting police inspectors by prison entrances was the only useful method of recognizing criminals who used false names. Those prisoners the officers didn't recognize, they greeted like old friends, attempting to fool them into giving away their true identities. Since a successful recognition won an inspector five francs, some inspectors conspired with prisoners to make false identifications, later splitting the reward with them. Bertillon had to copy even these false details into his records. The meaninglessness of his work disgusted him, but with his love in mind, he tolerated it.

Then, four months after he started, Bertillon struck upon an idea. He had thought of a way to uniquely identify criminals, even when they lied about their names, and it involved the

anthropometric measuring techniques he had learned as a child
from his father and grandfather. Many practical details had to
be worked out. He got permission from the governor of a local
jail to take experimental measurements from the prisoners
there, and by mid-August of 1879, Bertillon had confirmed that
his system could distinguish one person from another. He had
finally accomplished something. Bertillon sent an uninvited
report of his success to the head of the Paris police, Prefect Louis
Andrieux, but a month and a half later, he had heard nothing.

Bertillon desperately wanted permission to try out his new
ideas. On October 1, 1879, the day he was promoted from
assistant clerk to clerk, he sent a second report to the Prefect.
The rationale behind the system, Bertillon explained, was
Quételet's study of the human physique. Quételet, the statisti-
cian who published his study of French crime statistics in *A
Treatise on Man,* had also, in the same volume, reported his
investigations of the human physique. Analyses of the height
and chest measurements of army recruits had shown that men's
bodily dimensions were wide and varied. No two men's bodily
dimensions, Quételet concluded, were exactly alike.

Bertillon's system employed eleven separate measurements:
height, length and breadth of head, length and breadth of ear,
length from elbow to end of middle finger, lengths of middle and
ring fingers, length of left foot, length of the trunk, and length of
outstretched arms from middle fingertip to middle fingertip. The
odds of one particular measurement being exactly the same in
any two men were one in four, Bertillon's experiments had
shown. The chances of two men having, say, both the same
height and the same size foot, therefore, would be one in 4×4.
The chances of three common measurements were one in
4×4×4. And the odds of two men sharing all 11 bodily measurements were one in four to the eleventh power, or one in

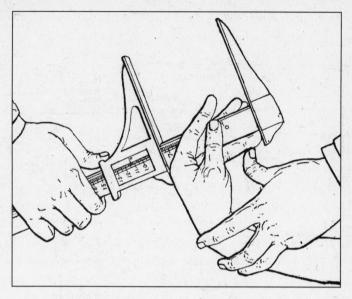

Length of middle finger

4,194,304. Augmented by photographs and very precise technical descriptions, what Bertillon called "spoken portraits," the measurements could distinguish one person from another.

But as Bertillon would later point out in his textbook, *Signaletic Instructions,* "The solution to the problem of judicial identification consists less in the search for new characteristic elements of individuality than in the discovery of a method of classification." How, in other words, in a vast collection numbering many thousands, could the file of a criminal, whose real name was unknown, be found with no other information than his measurements?

Bertillon solved the problem by defining, for each measurement, categories of small, medium, and large. A collection of, say, 90,000 file cards broke into groups of 30,000 according to

Length of outstretched arms

the three head-length categories. Further categorization, by head width, then subdivided the collection into groups of 10,000. This segmenting process, followed for seven of the eleven differ-ent measurements, resulted in groups of about forty-one files. Seven gradations of eye color, defined by Bertillon, further divided the cards until each classification contained between three and twenty cards. A criminal could be measured, looked up, and identified in a matter of minutes.

Bertillon's system was a triumph in logic and scientific method. His written report to Prefect Andrieux, however, was a flop. Uneven education had left Bertillon with poor writing skills, and he completely confused his ideas in an unclear and repetitious garble of words. Andrieux, who won his post through good political connections, knew little about police work and even less about mathematics. He had no patience for Bertillon's paper. He handed it for review to Gustave Macé, the head of the Sûreté.

Macé was the Sûreté's fifth chief since Vidocq's reign, forty years earlier. The former criminals who peopled the detective agency under Vidocq had been swept aside by inspectors with more respectable backgrounds, and the ranks of the brigade had swollen to several hundred. Professional detection was in its heyday. The dark times of the ordeal and the extraction of confession by torture were a distant memory. The newspapers entertained the public with long accounts of the stunning feats of logic through which detectives snared their quarry.

One sensational case had launched Gustav Macé's career. A restaurateur, fishing in his well for the source of the bad taste of his water, retrieved a severed human leg. Macé rushed to the restaurant in the Rue Princesse where, after dredging the well, he found another leg. Other body parts floated to the surface of the River Seine over the next few days, but not the head, so Macé faced the double challenge of identifying the victim as well as the murderer.

Macé's inquiries quickly led him to a tailor named Pierre Voirbo, who often stopped for water at the well where the legs were found. Voirbo, Macé discovered, had argued violently with his friend Monsieur Désiré Bodasse, who refused to loan Voirbo 10,000 francs for his wedding. When Macé found that

Macé fishes for the leg

Bodasse was missing from his apartment, he accompanied his aunt to the morgue, where she recognized his stockings and a scar on Bodasse's leg. Macé had identified the victim.

Voirbo, however, denied the murder, and Macé could find no trace of blood in his lodgings. Standing in the bedroom with Voirbo and two police officers, Macé seemed to have come to a dead end. Suddenly, Macé grabbed a bottle of water and gestured to pour it over the floor. If a corpse has been dismembered here, he said, blood must have collected in a pool at some place on the floor, depending on its slopes and dents. When I pour out

Finding where the blood collected

this water it will collect in the same place, he said. The water flowed towards the wall under the bed. Voirbo began shaking violently. Mop up the water and remove the floor tiles, Macé ordered. The floorboards beneath were covered with dried, caked blood. Don't continue, Voirbo cried. I am guilty. Case solved.

A fascination with their own successes made detectives like Macé arrogant. The pioneers of their profession, they liked to believe they had been specially endowed with bloodhound-like instincts and hawk-eyed powers of observation. They prided

themselves on practicality and disparaged the quasi-scientific
ideas thrown their way. Not that anyone could blame them,
since thus far pseudo-scientific means were useless at best. In
1829, a Frenchman named Barruel suggested he could identify
criminals by their odors, and later, Louis Figerio, Superintendent of the Royal Asylum for the Insane in Alessandria, Italy,
insisted he could tell a "born criminal" by the size of his ears.
Measuring the lengths of bones, Bertillon's idea, didn't seem any
less crackpot to Macé. He dismissed Bertillon's proposal uncer-
emoniously. The police, he wrote in a report to Andrieux, had
more important things to do than bother themselves with theo-
rists' experiments.

Andrieux called Bertillon to his office. "If I am not mistaken," he said, "you are a clerk of the twentieth grade and you
have been with us for only eight months, right? And you are
already getting ideas? Your report sounds like a joke." Bertillon
tried to explain but, stumbling over his own words, he was cut
off sharply. Don't worry your superiors any further with this,
Andrieux warned Bertillon. To make clear his point, he wrote to
Bertillon's father, threatening the son with dismissal if he med-
dled further. End of discussion.

In the same period, Andrieux also received a proposal to use
the nearly microscopic lines on the tips of the fingers to identify
criminals. It was from an obscure missionary doctor working in
Japan, Henry Faulds. Bone lengths? Finger lines? What's next?
Andrieux must have thought. Crackpot scientists were a plague.
He did not even bother to grace Faulds with a reply. His
response was typical. Faulds's carefully penned letters to police
chiefs around the world received not an inkling of interest. Nor
had Faulds's 1880 *Nature* article caused much of a stir. It only
inspired a couple of articles, a discussion at the 1881 International Medical Congress and, possibly, part of Mark Twain's
1883 book *Life on the Mississippi*.

"A Thumb-Print and What Came of It," the thirty-first chapter in Twain's book, told the story of Karl Ritter, whose wife and child were killed by a marauding Civil War soldier. Ritter set out to track down the murderer, with only a bloody thumbprint to identify his quarry. Posing as a palm reader, Ritter used fortune-telling as his excuse to examine the lines on the thumbs of every soldier he encountered. "These lines were never alike in the thumbs of any two human beings," Twain wrote. Though his fingerprint-based plot may have been inspired by Faulds's *Nature* article, it was small compensation for the doctor's foundering hopes of seeing his grand scheme adopted by the police.

Back in France, meanwhile, Bertillon's father had received Andrieux's threatening letter. He angrily sent for his problem son, who begged him to read his report. The senior Bertillon quickly recognized its value, and apologized for his reprimand. "I had no longer dared to hope that you would find your own path in the world," he said. "But this is the beginning of it. This is applied science, and it will mean a revolution in police work. I shall explain the matter to Andrieux. He must realize."

He didn't. Even the accomplished scientist could not convince Andrieux of the system's merits. Only after the Prefect resigned, three years later, did Bertillon get the chance to prove himself. In mid-November 1882, Jean Camecasse, the new Police Prefect, called the young man to his office. He offered Bertillon two assistant clerks to help set up his system. Camecasse gave him three months to identify the first recidivists and prove his scheme. It was an unfair challenge. How often does a criminal get caught, charged, sentenced, serve his sentence, come out, and get arrested again in only three months? Nevertheless, Bertillon accepted.

He worked obsessively, measuring prisoners as fast as they were arrested. Each night, he carried a portfolio full of data to the apartment of an Austrian language teacher, Amélie Notar, a woman friend whom he had met in the street (it was she, not the anonymous Swede, whom he would eventually marry). Because Bertillon did not trust his assistant clerks, Amélie transcribed his data on to cards for him. By the beginning of January 1883, his files contained 500 cards. By mid-February, with his three months nearly up, he'd collected 1,800 cards. Still, he had made no identifications. His migraines and nosebleeds got worse.

On February 20, as Bertillon measured the day's sixth "Dupont"—then a favorite nom de guerre of criminals trying to hide their true identity—a spark of recognition lit in his mind. "Dupont" looked familiar. Bertillon checked his files for the man's measurements, and found them. "Dupont's" real name was Martin, and he'd been arrested two months earlier on December 15. He had stolen milk bottles. When Bertillon confronted him, the man confessed his true identity.

On February 21, 1883, Paris newspapers published excited accounts of Bertillon's first identification. Prefect Camecasse, basking in the positive publicity, gave Bertillon more clerks, an office of his own, and indefinite leave to continue his system. Bertillon's father, who was mortally ill, died a happy man. The world had accepted its first scientific method of identification, and the credit belonged to his once wayward son.

. . .

While Bertillon built on his successes, Faulds was back in Scotland, recovering from a bout of malaria. Doctors had sent him home from Japan with his family in 1882, forcing Faulds to postpone a long-term study to establish the persistence of finger-ridge patterns over time. Faulds later came under fire for not completing the experiment and conclusively proving long-term persistency, but his life leapfrogged from one misadventure to the next, and his study fell by the wayside.

In 1884, recovered from his illness, Faulds sailed back to Japan. His wife and children, following from Scotland some months later, were shipwrecked off the coast of Tokyo. They survived, but Faulds's wife, he wrote in his personal notes, "received so severe a shock that under medical orders I had to return finally to England in the late autumn of 1885." By some accounts, his wife, Isabelle, had taken to drink. Back in Scotland, he wrote, "My father then took ill and died, and with a sick wife and young family I had to look for an immediate livelihood, and had as hard a time—I should think—as any professional man experiences in this country today."

Faulds was like a prize racehorse stuck in his starting box while his competitors galloped ahead. At every turn, some new problem frustrated his attempts to promote his fingerprint system, giving Bertillon's anthropometry a huge head start. Only after he moved to London in 1887, four years after Bertillon's first identification, did Faulds earnestly begin a two-year struggle to get Scotland Yard to adopt his fingerprint system of criminal identification.

The junior detectives Faulds first approached at the Yard initially greeted Faulds's ideas with skepticism. According to Faulds's book *Guide to Finger-Print Identification,* one young officer asked "whether he was really required to believe that a tiny patch of skin like that of a finger-tip could contain a variety of lines enough to help in identification." Faulds pointed to a map of London that hung on the wall, and circled a fingertip-sized patch. He showed the detectives that the patch contained enough street detail to determine from what city it had been taken, and that a similar idea applied to finger-ridge detail. He sold them on the idea, but a tangle of bureaucratic red tape prevented further progress. Faulds wrote to the Yard asking to speak with a more senior detective.

In 1888, Chief Inspector John Bennet Tunbridge of the CID

visited Faulds at home. Two or three years later, Tunbridge
would handle the famous case of Thomas Neill Cream, a serial
killer who poisoned at least six women, and six years after that,
he would take the post of commissioner of the New Zealand
police. Faulds could not have hoped for a more senior inquisitor
into his system. He demonstrated to Tunbridge his method for
inking and printing all ten fingers in serial order on a specially
prepared form, and explained the classification system he had
come up with to allow easy retrieval of criminals' files. The classification system categorized each fingerprint according to its
graphical elements, an idea Faulds borrowed from the way
Japanese writing symbols were organized in dictionaries.

Each of Japan's 30,000 symbols, or ideographs, was con-
structed around one of 212 primary elements, known to western
lexicographers as a radical. In Japanese dictionaries, ideographs
based on the same radical were grouped together, and the groups
were ordered according to the number of brush strokes contained in the radical. Within each radical category, ideographs
were subclassified by the total number of brush strokes contained
in the whole character. A character could then be quickly found
in a dictionary by determining its category and subcategory.

Similarly, Faulds's system for classifying the "hieroglyphics
of the human fingers" enabled a criminal's file to be found easily
in the corresponding category of a vast fingerprint collection.
The "radical" of the fingerprint was the dominant feature at the
core of its pattern. Faulds categorized the cores by feature, each
represented by a Roman consonant. The letters V and W, for
example, represented counterclockwise or clockwise whorls or
spirals. M and N represented figures "somewhat resembling
mountain peaks, M signifying an outline like that of a typical
volcanic peak, while N, though similar, ends in a rod-like form,
as of a flag-staff on a mountain top."

"We have thus," Faulds wrote, "with the use of consonants alone, built up a kind of skeletal system, and we have now but to add the vowels to make these dry bones speak." He used vowels to represent the additional elements that modified the core, like the extra brush strokes that modified an ideograph radical. The letter A, for example, indicated that the interior of the core's loop or whorl, potentially enclosing further detail, was empty. E indicated that the interior enclosed a group of not less than three detached lines. Thus, a consonant and one or two vowels—a syllable—represented each fingerprint. A five-syllable word described the entire hand's five fingerprints. The system classified a criminal's entire series of ten fingerprints under the two five-syllable words that represented each hand.

Faulds explained all this to Tunbridge. He even offered to set up a model bureau at the Yard at his own expense. Tunbridge was unconvinced. Unlike Bertillon's anthropometry, fingerprints weren't rooted in any known science. Nor, like Bertillon, did Faulds have a well-known, socially connected scientist father to vouch for him. "At the close of our long interview," Faulds reported, "[Tunbridge] told me he was disposed to think the method would be rather delicate for practical application by the police, and that fresh legislation would be required before any beginning could be made."

Bertillon's system, meanwhile, had gathered momentum. After his first identification in February 1883, Bertillon took only about a month to make a second. In the following three months he got three more; the three months after that he got fifteen; and in the remainder of the year he identified twenty-six prisoners. In 1884, his first full year of operation, Bertillon identified 241 recidivists—all of whom detectives had failed to recognize. His success in recognizing habitual criminals facilitated the administration of France's new Relegation Law, which man-

dated lenient treatment for first-time offenders and greater punishment for repeaters.

Though Gustave Macé, Bertillon's oldest critic, had resigned
as chief of the detective branch in April 1884, members of his
old department still refused to take young Bertillon seriously.
They resented his rapid success and made every attempt to
block the expansion of his system. They frequently humiliated
him by calling him to the morgue to identify murdered men and
ridiculed him when he turned green.

Finally, as a joke, they called on him to identify a bloated
and stinking body found floating on the Marne. Bertillon measured it and, because of its especially large head, easily found a
matching card in his files. The body belonged to a man who,
twelve months previously, had been convicted of violent assault.
Following Bertillon's lead, police investigations discovered that
the man had been missing from his home for two months.
Bertillon received a new respect from the detective bureau. The
department's resistance to his system dissolved.

In 1885, the French national prison system adopted anthropometry and the papers made Bertillon a hero. "Young French
Scientist Revolutionizes the Identification of Criminals" and
"Dr. Bertillon's Ingenious Method of Measurements," read the
headlines. Bertillon gained international renown, and his system
spread throughout the continent. In 1887, it crossed the Atlantic
to the United States and Canada. Anthropometry still had not
jumped the Channel, and as long as Britain had adopted no
scientific system of identification, there was still some lingering
hope for Faulds's fingerprinting.

By 1888, however, when Bertillon was installed as Chief of
France's newly established Service of Judicial Identity, British
officialdom had begun to catch Bertillon's scent. Many distinguished visitors from around the world had paraded through
Bertillon's office, but that year saw his first distinguished British

visitor: the famous scientist Francis Galton, the cousin to whom Charles Darwin had sent Henry Faulds's letter eight years earlier.

Galton's visit to Bertillon's laboratory was not spurred by interest in identification. His fascination was with the application of Darwin's theories to create a master race of men. He hoped to identify Britain's most able-bodied and nimble-minded and to mate them the way a horse breeder creates winners of the Grand National. To do this, he needed to identify which members of the population had the greatest genetic potential. Galton had decided bodily measurements might be one way to accomplish this and, for this purpose, four years before he visited Bertillon in Paris, Galton had opened an anthropometric lab of his own.

Anthropometry would ultimately be just one in a number of methods Galton tried using to determine hereditary potential. He also considered analysis of facial characteristics, recording the success of people's forebears, and, indeed, fingerprints. Serendipitously, therefore, Galton became an expert in the very methods police considered for the identification of criminals, and Scotland Yard would eventually come to him for advice. Galton would have great influence over Britain's choice of an identification system. It was as though, when Galton visited Bertillon's laboratory in 1888, fate dropped the future of Henry Faulds and his fingerprint idea right in Galton's lap.

Six

A Biological Coat of Arms

Francis Galton's arrival on Henry Faulds's stage was like the antihero's entrance in a tragically ending play. Raised in a high-society family, with all the wealth, prestige, and connections that entails, Galton took his privilege for granted. Moreover, he was notorious for using his status against those with fewer advantages. He believed that those of upper-class birth were by nature superior to the lesser-born. Thus, when he benefited from the hard work of a member of the lower classes, he felt entitled to do so without credit or acknowledgment. Worse yet, there was no recourse. Those who dared to oppose him learned that he was, by all accounts, that dangerous breed of dog who bites before even bothering to growl.

Born in 1822 in Sparkbrook, near Birmingham, Galton was fathered by a wealthy banker, who was in turn the son of a prosperous gun manufacturer. Galton's mother was the daughter of the scientist and doctor Erasmus Darwin, Charles Darwin's grandfather. By an interval of six years, Galton was the youngest of their eight children. Galton's sisters fought for the chance to play with baby Francis, and his mother, to keep things quiet, hung her watch on the wall to time their fifteen-minute turns. Galton was, as his biographers are fond of saying, "indulged." He was also intellectually gifted.

By two and a half, Galton could read; by four, he hoarded pennies for university; by five, he had read Homer's *Iliad*. But while family mollycoddling encouraged advancement at what he liked, it left him disinclined to apply his mind to what he did not.

Galton performed poorly in grammar school; he hated the classics. At sixteen, bored by his medical training, he amused himself by sampling each medicine in alphabetical order, until croton oil caused the contents of his stomach to head for both exits, and banished him to the toilet for hours. After persuading his father that he needed a break from medicine, he studied mathematics at Trinity College, Cambridge, and received a degree without honors. Galton ascribed his failure to overwork. During the same period, however, he invented a new type of oil lamp, a lock, a balance, and a kind of steam engine.

When his father died in October 1844, Galton abandoned his medical training once and for all. He lived on his inheritance and drifted aimlessly from one leisure pursuit to the next, traveling and hunting. Finally, in April 1850, perhaps inspired by his cousin Charles Darwin's HMS *Beagle* exploration, Galton set himself a worthy goal. He sailed south to explore an as yet uncharted part of South Western Africa (now Namibia). Eighty-six days later he landed, with his expedition partner, Karell (also known as Charles) Andersson, at Cape Town.

Their expedition was beset by difficulties. Midday temperatures soared to 157 degrees Fahrenheit in the sun, and a war raged between the Damara and Namaqua tribes on the land Galton hoped to cross. Their vicious fighting left children with gouged eyes and women with their feet severed for their copper anklets. Galton brokered a peace between the tribes, and pushed his expedition forward. Ten Europeans, eighteen Africans, and a caravan of sixty oxen and mules labored in the heat. Galton, all the while, kept copious journals and charts.

Francis Galton

Back in England in 1853, Galton's and Andersson's explorations together won Galton a Royal Geographical Society membership and one of its two annual Gold medals. Karell Andersson, however, received no award. When he arrived home penniless, he turned to Galton for help. Instead of offering thanks for Andersson's bushwhacking expertise and the awards it had helped him win, Galton wrote Andersson a sanctimonious letter, refusing assistance. Andersson's destitution, Galton told him, was his own fault. He should have worked his way home from Africa as a sailor, and saved his passenger fare. "A fatal pride," Galton wrote, "has placed you in a very false position." Galton, at that time, lived comfortably on an inheritance of £26,000—several million dollars, by today's standards.

Galton increased his prestige that year when he published his travel account *Tropical South Africa*. By 1856, he was married, and had taken his place as a Fellow of the Royal Society and a member of the prestigious Athenaeum Club. During the next ten years, he participated actively in Royal Geographical Society affairs, published another book, *The Art of Travel*, and made a number of contributions to meteorological theory. For the most part, however, he fell back into the unfocused ways of his youth. He treated with ingratitude and callousness both those who helped him win his prestige and those who through hard work had won acclaim of their own—such as Andersson, and, eventually, Faulds.

In 1871, Galton's lack of sympathy for his perceived inferiors gained widespread attention. Galton chaired a British Association meeting crowded with spectators eager to hear the exciting tale of Henry Stanley, the British-American explorer and journalist. Stanley had won an undeclared race with the Royal Geographical Society expedition to rescue the famous explorer Dr. David Livingstone, who was stranded in western Africa. Stanley's victory embarrassed the Society. Galton was outraged. During the question-and-answer period, he accused Stanley of sensationalizing his account, and referred to rumors of his illegitimate birth. In front of a crowd of 3,000, Galton essentially called Stanley a bastard.

Far from ever apologizing to Stanley for his public outburst at the British Association meeting, Galton went on to write to Clements Markham, president of the Royal Geographical Society, suggesting that the public and the Queen should be informed of the circumstances surrounding Stanley's birth. The Queen, however, received Stanley and thanked him for his services. Markham later wrote what he thought of Galton after the Stanley affair: "He was essentially a doctrinaire not endowed with much sympathy. . . . He could make no allowance for the failings of others and he had no tact."

Other people's success aroused a venomous jealousy in Galton, particularly when they achieved it by hard work instead of birthright. For much of his life, Galton had ridden on his forebears' coattails. He lived in the shadow of his family's great men. He had made only a start, with his Africa explorations, at building his own reputation, and felt threatened when someone with lower social standing superseded his prestige. Handicapped by the legacy of a spoiled childhood, he had still not adjusted to a world that prized what an individual achieved more than what he had been given.

Galton wrote, "I have no patience with the hypothesis . . . that babies are born pretty much alike, and that the sole agencies in creating differences between boy and boy, and man and man are steady application and moral effort. It is in the most unqualified manner that I object to pretensions of natural equality." For his ideas on heredity and human breeding, Galton later coined the term "eugenics." Later, in 1933, Adolf Hitler's government would appropriate the term when it passed the Eugenic

Sterilization Law, which ordered compulsory sterilization of all German citizens with presumptively inherited afflictions.

Galton's social elitism explained his outrage at Henry Faulds's eventual claims for the respect and credit Faulds was due. Each man was everything the other was not. Galton was rich; Faulds was poor. Galton did not receive his physician's licentiate; Faulds did. Galton was an atheist; Faulds was religious. Galton resided at the center of society; Faulds was an outsider. Most important, much of what Francis Galton got in life, he never had to work for, while much of what Faulds worked for, thanks in part to Galton, he never got.

.   .   .

If Galton ever needed proof of his ideas of the genetic superiority of some people over others, it had come, in his mind, with the 1859 appearance of his cousin Darwin's theory of evolution. Here was proof that his genetic heritage, in spite of his sometimes dilatory existence, exalted him. The children of the intellectually and physically well-endowed were naturally superior, his logic went, and this therefore took precedence over his lack of accomplishment. Ironically, launching himself on a quest to prove this fact would finally make him an accomplished and hard-working scientist in his own right.

His work on heredity began when he acknowledged one of his own hard-to-admit weaknesses. Galton's fruitless attempts to have children with his wife forced him to notice that childlessness ran in both their families. He had also noticed, at Cambridge, that academically talented students tended to come from academically talented parents. If fertility and intelligence were passed from parent to child, Galton thought, then physical prowess and creative talent must be transmitted, too. The deliberate coupling of the country's smartest and ablest, according to

Galton's interpretation of Darwin's theory, would lead, over several generations, to a physically and intellectually improved population.

Another deduction Galton made from Darwin's theory was that God did not exist. He wrote in a letter to Darwin that *Origin* had driven away "the constraint of my old superstition as if it had been a nightmare." Galton even tried to prove scientifically that prayer did not work. He compared the life spans of eminent clergymen and doctors and found, on average, that doctors lived about six months longer. Galton concluded that "the prayers of the clergy . . . for recovery from sickness, appear to be futile."

In contrast to the thinking of men like Darwin's colleague Asa Gray and Henry Faulds, Galton rejected the idea that a cosmic puppeteer pulled at man's evolutionary strings. Evolution was not divinely directed, and man might just as easily evolve backward towards the apes as forward into the image of his once-fancied Creator. Man's true religious duty, therefore, was not to any church. Man's allegiance should be to the deliberate and systematic forward evolution of the human species.

In 1869, Galton published *Hereditary Genius,* his first book on the subject of heredity. He had studied the family trees of eminent men and found that, more often than not, their forebears had also distinguished themselves. This vindicated, in Galton's mind, his insistence that the intellectual talents of men were inherited and innate, not developed. The results cemented his place on the nature side of the nature-versus-nurture debate. "As it is easy . . . to obtain . . . a permanent breed of dogs or horses gifted with peculiar powers of running . . . ," Galton wrote in *Hereditary Genius,* "so it would be quite practicable to produce a highly-gifted race of men."

His first step in creating his "highly-gifted race" was to determine the average levels of physical and mental prowess possessed by the nation—the sort of evolutionary starting block from which his eugenic program would begin. Good breeding stock could then be selected by comparison of their measured abilities with the evolutionary average. Galton embarked on the immense task of collecting the physical and psychological data that would allow him to compile this evolutionary average. He also began a hunt for a sort of biological coat of arms, a physical feature that distinguished the ablest, smartest, and fittest—the breeding stock—from the hoi polloi.

Galton first hoped that the face might contain such a mark. To test his hypothesis, he obtained photographs of criminals serving long sentences and sorted them according to crime—murder, forgery, and sex offense. Each crime, Galton reasoned, correlated with a particular psychological sort. He exposed single photographic plates to a series of eight images of criminals of a single type, superimposing them. These "composite photographs" resulted in the visual average of the criminals' faces. Individual peculiarities were lost while common aspects were retained and reinforced. Disappointingly for Galton, the results were indistinguishable from results obtained from law-abiding citizens. "The special villainous irregularities . . . have disappeared," Galton wrote of the composites, "and the common humanity that underlies them has prevailed." He would have to look elsewhere for his biological coat of arms.

Meanwhile, his collection of physical and psychological data accelerated when he opened in 1884 an anthropometric laboratory at the International Health Exhibition (he later moved it to the Science Museum at Kensington). Unlike the lab where Bertillon had made his first identification one year earlier, Galton's had no connection with law enforcement. Its sole purpose was to collect more information.

After a doorkeeper collected a threepence admission fee, the superintendent guided each patron along a long table of instruments. He measured the patron's weight, sitting and standing height, arm span, breathing capacity, strength of pull and squeeze, force of blow, reaction time, keenness of sight and hearing, color discrimination, and judgments of length. At the end, each patron received an impressive-looking certificate recording the results of the examination, and Galton, during the laboratory's existence, added 10,000 sets of data to his evolutionary studies.

The 1888 press reports of the launch of Alphonse Bertillon's new Service of Judicial Identity, the location of the world's other large-scale anthropometric laboratory, aroused Galton's curiosity enormously. Galton visited Bertillon's laboratory during a short vacation with his wife to Paris. Though Bertillon's measurement methods interested Galton more than their application to criminals, Galton's anthropometric expertise gave him the authority to review Bertillon's system of criminal identification before the British scientific community. On Galton's return from Paris, the Royal Institution invited him to give a Friday evening lecture on the subject.

Galton couldn't bring himself to selflessly glorify someone else's work. He decided to give the lecture the more general title of "Personal Identification and Description," and to add research results of his own. Some of the methods he had considered for discovering hereditary potential, such as the careful measurement of facial characteristics, could be applied to identification. He also cast around for other potential identification systems. While preparing for the lecture, Galton much later explained in his book *Memories of My Life*, "the fact occurred to my recollection that thumb-marks had not infrequently been spoken and written about, so I inquired into their alleged use."

Eight years earlier, Charles Darwin had passed Henry Faulds's letter on the subject of fingerprints to Galton, and Faulds had published his letter in *Nature*. More recently, Galton shared a train carriage with William Herschel on the way to the annual meeting of the British Association for the Advancement of Science. Since his retirement from India in 1878, Herschel had taken a theology degree at Oxford, and actively participated in the Temperance Society, campaigning for the prohibition of alcohol. He also maintained his family's interest in science, the reason for his attendance at the British Association. During their train ride together, Herschel excitedly recounted to Galton his use of fingerprints in India.

Galton scoured the scientific literature for further discourse on fingerprints. By far, the two most substantial contributions were Faulds's and Herschel's 1880 letters to *Nature*. Herschel's letter discussed his use of fingerprint signatures in India, and his observations that individuals' fingerprints did not change over time. Though the long-term persistence of fingerprints was important to their use in identification, this one original contribution in Herschel's letter did not hold a candle to the rich and varied contents of Faulds's.

Faulds had written not only about the application of fingerprints to criminal identification, but about their possible use in the study of man's origins and in ethnology. Evidence that fingerprints might be useful in these fields, Faulds wrote, included the fact that the fingerprints of monkeys were similar to humans', and that patterns of parents were often similar in their children. "The dominancy of heredity through these infinite varieties is sometimes very striking," Faulds wrote.

That was all Galton needed to kindle his interest. If the patterns were hereditary, then perhaps elements in the patterns were related to certain family characteristics, like strength, intelligence, or dexterity. Perhaps fingerprints were the biological coat of arms Galton had been searching for. Faulds's fingerprints finally had the attention and interest of a prestigious scientist with the influence to further his fingerprint studies. It should have been the great turning point in Faulds's life.

But Faulds, an obscure Scottish doctor, was completely unknown to Galton. William Herschel, on the other hand, was very familiar to him. Galton had used the Herschel family as an example to support his ideas on the heredity of talent in his book *Hereditary Genius*. In 1860, he had learned the operation of a certain astronomical instrument from William's father, Sir John, in preparation for a journey to Spain to observe the solar eclipse. Galton was probably also enamored of the hereditary Herschel family title. Though Faulds had published a far more significant and valuable contribution on fingerprints, Galton, ever the elitist, preferred associating with Herschel.

On March 1, 1888, Galton wrote to Herschel, telling him of his interest and asking Herschel to forward fingerprint samples: "It would save me much trouble if I started from the high level of your large Indian experience." Herschel responded straight away, sending Galton his own fingerprint retaken over a period of twenty-eight years. "You can judge for yourself whether the persistence of the ridges is not astonishing," he wrote. He also included instructions for taking finger impressions using printer's ink spread thinly on a piece of tin.

Galton added thumb impressions to the other data collected at his anthropometric lab. He regularly corresponded with Herschel, hungrily picking his brain for anything more he could tell him about fingerprints. With Galton's encouragement, Herschel came to consider that he, not Faulds, deserved to be known as the true inventor of the fingerprint idea. Herschel had, his reasoning went, impressed Konai's hand in 1858, twenty years before Faulds became interested in the subject. That Faulds had developed the idea much further than Herschel had and had been the first to publish the idea was incidental.

That year, 1888, the same year Faulds tried to convince Scotland Yard to adopt the fingerprint system, Herschel and Galton privately agreed to promote themselves as fingerprinting's pioneers, according to a letter Herschel wrote to the *Times* many years later. In return for Herschel's assistance, Galton would tell the world that Herschel had put the "finger-print system into full and effective work . . . as early as 1877, after some 20 years' experimenting for this one definite purpose." In fact, "full and effective work" hardly described Herschel's early use of fingerprints. He had never used fingerprints to identify criminals who lied about their identity or who had left their marks at crime scenes. He had merely used fingerprints as a form of signature on deeds and jailers' warrants.

Promoting Herschel as the originator of fingerprinting had its advantages for Galton. Claiming he took his lead from Herschel, Galton need never credit Henry Faulds with the ethnographical and criminal identification ideas Galton intended to develop. He could say that they were the natural extensions of Herschel's ideas, which he pursued with Herschel's blessing. By the version of events they agreed upon, Herschel would be the system's originator, and Galton its developer. They left Henry Faulds out in the cold.

. . . .

Galton delivered his lecture "Personal Description and Identification" at the weekly evening meeting of the Royal Institution on Friday, May 25, 1888. He described Bertillon's anthropometric method and several methods of his own conception, including the measurement of facial features and the cataloguing of minute particulars of a person's body, such as the markings of the iris of the eye and the convolutions of the external of the ear.

"But by far the most beautiful and characteristic of all superficial marks," he said, "are the small furrows, with the intervening ridges and their pores that are disposed in a singularly complex yet regular order on the under surfaces of the hands and feet." He recounted Herschel's use of fingerprints to put "an end to disputes about the authenticity of deeds," explained Herschel's methods of taking prints, and generally gave the impression that Herschel had been the first to publicly intro-duced the fingerprint idea. He directed his listeners to Herschel's *Nature* paper, and referred, in passing, "also to a paper by Mr. Faulds in the next volume."

By citing Herschel's and Faulds's publications in reverse order, Galton gave the appearance that Herschel's came first. Whether this subtle falsehood was a mistake or a deliberate attempt to rewrite history, only Galton knew. But the effect, later reprinted in an article in *Nature,* was clear. It effectively robbed Faulds of his priority of publication and his rightful place as the first announcer of a new discovery. Galton, mean-while, continued to build on both the identification and ethno-logical fingerprint work that the doctor had begun in Japan.

Before long, at his Science Museum laboratory, Galton had collected the right and left thumbprints of 504 individuals. He photographically enlarged each print by a factor of two and a half times, obtaining a deck of 1008 playing-card–sized thumbprint samples. He began trying to lump the cards together by pattern type. Galton hoped to discover a relationship between the most fundamental pattern types and certain physi-cal or mental characteristics. Perhaps the most common thumb pattern related to the most average people. Perhaps particularly rare patterns indicated particularly rare forms of intellectual tal-ent or physical prowess.

Faulds, in his *Nature* paper, had mentioned two categories of fingerprints, namely loops and whorls. Loops were formed when the ridges, running from one side of the fingertip to the other, turned back on themselves. Whorls were formed when the ridges turned through at least one full circle. But two pattern types did not provide anywhere near enough discrimination for Galton's purposes. Much finer distinctions between pattern types were contained in Faulds's unpublished classification system, based on Japanese ideograms. The doctor would have surely explained the system to Galton if he had been contacted, but Galton insisted on finding his own way of lumping together the patterns.

At first, Galton attempted to sort his deck of thumbprints according to the nine pattern types noted by Jan Evangelista Purkyně, the Czech physician who had discussed the ridges on human hands in his 1823 thesis on the skin. But Galton could not force his thumbprint collection to divide itself neatly into Purkyně's categories. He next attempted to divide his collection into sixty pattern classifications of his own devising. His problem now was that many of his thumbprint samples fell easily into more than one classification.

Finally, after eliminating any pattern types that could be turned back to front or upside down to make another, he found himself back where he started, with Faulds's loops and whorls. He next divided loops into two categories, outer and inner, depending on whether the base of the loop opened inward toward the rest of the fingers or outward toward the thumb. To these three categories Galton also added arches, formed when the ridges run from one side to the other of the bulb of the digit without making any backward turn or twist. Reluctantly, he admitted that these four unsophisticated classifications represented fingerprints' most fundamental pattern types. Without being able to discriminate more than four pattern types, Galton could not correlate them to other physical or mental characteristics. The thumbprint would not work as Galton's biological coat of arms, but maybe, Galton thought, a full set of fingerprints would.

In 1890, Galton began collecting the prints of all ten digits—another idea originally contained in Faulds's *Nature* paper. He classified his ten-print collection by labeling each finger or

Arch Loop Whorl

thumb, except the forefinger, with the letters *a, l,* or *w* to indicate arch, loop, or whorl. Forefingers also carried the *a* and *w*
labels, but loops were further distinguished with an *i* or an *o* to
indicate inner or outer. Each fingerprint set was then indexed by
its corresponding list of letters. Galton would eventually suggest
this unsophisticated system to Scotland Yard for the purposes of
classifying a criminal fingerprint register.

Galton's more interesting contribution was his method of
distinguishing fingerprints that contained similar patterns. The
general fingerprint patterns of twins, for example, were often
identical. But Galton had noticed that fingerprint ridges did not
proceed across the fingertips in unbroken lines. They often
stopped abruptly, split, contained enclosures, or connected with
other ridges. The arrangement of these ridge details were never
repeated in a print from two different fingers, not even in twins.
Identification of one fingerprint with another, Galton realized,
should always be made by comparing their ridge detail or fingerprint minutia (known later as points of comparison or identification). He used this comparison of ridge detail to confirm
Herschel's observations of fingerprint permanence.

While in India, as a hobby, Herschel had collected the fingerprints of friends and family. At Galton's request, in the early

RIDGE CHARACTERISTICS	
	RIDGE ENDING
	BIFURCATION
	LAKE
	INDEPENDENT RIDGE
	DOT or ISLAND
	SPUR
	CROSSOVER

Types of ridge minutia

1890s, Herschel collected repeat impressions from fifteen individuals, including himself. Intervals between each individual's prints reached variously from babyhood to boyhood, from childhood to youth, from youth to advanced middle age, and from middle age to old age, allowing Galton to examine fingerprint persistence over the whole of man's life span. Galton examined the fingerprint minutia for changes that might have occurred over the years and found that they remained consistent. A person's fingerprints, he confirmed, would identify him for life.

After three years of researching and writing about fingerprints, Galton became sufficiently confident in the method to say that it would indeed form the basis for a reliable system of identification. In an 1891 article for the magazine *Knowledge*, he wrote: "I look forward to a time when every convict shall have prints taken of his fingers by the prison photographer, at the beginning and end of his imprisonment, and a register made of them; when recruits for either [military] service shall go through an analogous process; when the index-number of the hands shall usually be inserted in advertisements for persons who are lost or who cannot be identified, and when every youth who is about to leave his home for a long residence abroad, shall obtain prints of his fingers at the same time that the portrait is photographed, for his friends to retain as a memento."

A year later, in 1892, he published his comprehensive book *Finger Prints*, in which he discussed everything from his research into the history of man's interest in fingerprints to his attempts to find a correlation between fingerprint patterns and race, class, and intellectual prowess. Though Galton had appropriated Faulds's ideas without giving him credit, Galton nevertheless did move fingerprint science a long way forward. Even Faulds, not wishing to deny credit the way Galton denied it to him, would later acknowledge Gatlon's good work. Galton provided the systematic proof of its scientific basis, without which the fingerprint system would never have achieved general acceptance.

Sadly for Galton, however, fingerprints did not turn out to be the biological coat of arms he hoped for. They did not correlate in any way with physical or mental characteristics. He found that the "English, Welsh, Jews, Negroes and Basques may all be spoken of as identical in the character of their finger prints." Nor did a difference exist in the fingerprints of scientists, artists, men of culture, or the "lowest idiots in the London district." Galton so despaired of fingerprints' lack of relevance to his eugenic studies, that, initially, he did not bother promoting their official use as a method of identification. That task fell to William Herschel.

Galvanized by the publication of Galton's *Finger Prints*, Herschel wrote excitedly to an old colleague in India, Henry Cotton, who was now the Chief Secretary to the Government of

Bengal. He suggested the application of fingerprints to various offices in Bengal and included a copy of *Finger Prints*. Cotton, in turn, distributed Herschel's suggestion to the relevant departments. Meanwhile, that same year, across the globe in Argentina, a detective was already working on the world's first documented murder case involving a fingerprint.

Seven

Britain's Identity Crisis

Late one evening, in the poverty-stricken outskirts of a small Argentinian coastal town called Necochea, Francisca Rojas ran screaming from her hut and into her neighbor's fifty yards away. "My children," she cried. "He killed my children." The neighbor sent his son to fetch the police chief, while he and his wife tentatively ventured into Rojas's hut. They found her children, a boy of six and a girl of four, dead in their blood-drenched bed. Their heads had been smashed. The date was June 29, 1892, the same year Francis Galton published *Finger Prints*.

When the Necochea police chief arrived, the distraught mother accused a local ranch hand named Velasquez of killing her children. Velasquez was madly and jealously in love with Rojas. Earlier that afternoon, he had proposed to Rojas, but she said she would never marry him. She loved someone else. Velasquez raged. He threatened to kill her children. When Rojas returned from work that evening, she told the police chief, Velasquez came rushing out of her hut. Her children were dead.

The police chief immediately began a manhunt for Velasquez. He was so convinced by Rojas's moving and disturbing story that he never bothered to examine the murder scene for clues or weapons. After he found Velasquez and arrested him, he tried to beat a confession out of him. Velasquez admitted threatening Rojas, but no matter how much the police chief tortured him, he denied ever touching the children. The police chief kept pushing for a confession. He tied Velasquez down in a lit room beside the children's mutilated corpses, forcing Velasquez to look at them through the night. Even after this, and further torture, Velasquez denied the crime.

The police chief, meanwhile, discovered that Rojas had neglected to reveal that her other lover, whom Rojas loved deeply, hated her children. The lover had in fact told Rojas that he would marry her if only it weren't for her two brats. Suddenly, it dawned on the police chief that Rojas could have murdered her own children for this man, but he had no proof. In his haste to arrest Velasquez, he had left Rojas ample time to destroy any clues and dispose of her murder weapon. His only hope was to get her to confess. In desperation, he crept around her hut one night, moaning and wailing as though he were an avenging spirit come to punish the infanticide. Rojas was unmoved, and the chief was back where he started.

When the crime remained unsolved after a week, the chief called for assistance from the larger police force in La Plata, capital of the Buenos Aires Provincia. Police Inspector Eduardo M. Alvarez arrived in Necochea on July 8. By process of elimination, he quickly confirmed the chief's theory. Several witnesses had been with Velasquez on the night of the murders, though Velasquez, who was not especially bright, had never thought to mention this exonerating evidence. Rojas's other lover was nowhere in the vicinity at the time of the murder. That left only one suspect: Rojas.

Alvarez went to Rojas's hut and carefully scoured it for clues she might have overlooked. After several frustrating hours, the sun fell on the wood of the half-opened door, and Alvarez noticed a gray-brown spot on the wood. It was a smear of blood left behind by a thumb. Examining it, Alvarez suddenly recog-

nized the type of finger-ridge detail he had seen at the La Plata
Office of Identification and Statistics. The head of the Office,
Juan Vucetich, had used fingerprints there since the previous
July, 1891.

That month, the La Plata chief of police, Guilermo Nuñez,
ordered Vucetich, his subordinate, to establish an identification
bureau based on some articles about anthropometry he had seen
in French technical journals. Vucetich, a thirty-three-year-old
Croatian emigré, got straight to work taking experimental measurements from prisoners. A few days later, Nuñez dropped by
Vucetich's office with another article, this time from the May 2
issue of *Revue Scientifique*. The article reported on Francis Galton's experiments with fingerprints and explained their potential
use in identification.

The fingerprint conception fired Vucetich's imagination. He
immediately added the ten fingerprints to the measurements he
took from arrested men. On September 1, 1891, after experimenting for less than two months, he opened his Office of Identification and Statistics. On his first day of operation, having
already worked out his own crude system of fingerprint classification, he identified twenty-three recidivists from their fingerprints. But a year later, when Alvarez discovered the bloody
thumbprint on the door of Rojas's hut, Vucetich had yet to convince his skeptics in the government of the value of his system.

Alvarez cut out the section of Rojas's door containing the
thumbprint and took it to the police station in Necochea. He
ordered Rojas's arrest. With an ink pad borrowed from the
Necochea police chief, he took impressions from her thumbs
and compared them under a magnifying glass with the bloody
mark on the door. Rojas's right thumb matched perfectly. As
Alvarez explained to Rojas how fingerprints identify the fingers
they come from, she started shaking. When she saw for herself
that the patterns under the magnifying glass were the same, she

became hysterical, and, eventually, confessed everything. Because they stood between her and the man she loved, she had crushed her children's heads with a stone. She threw her murder weapon down a well and carefully washed the blood from her hands. She had forgotten that she had touched the door.

Back in La Plata, Vucetich believed that the solution of the Rojas murder completely vindicated his decision to pursue fingerprints instead of anthropometry. "I hardly dare believe it," he wrote to a friend, "but my theory has proved its worth. No doubt my opponents will call it fortuitous. But I hold one trump card now, and I hope I shall soon have more." Vucetich's superiors were less impressed by the solution of the Necochea murder than he hoped, and in 1893, they ordered him to put aside his fingerprint work and to concentrate on the internationally established methods of Bertillon. The following year, however, in 1894, a new police chief at La Plata allowed Vucetich to revive his fingerprint system.

By this time, Vucetich had much refined his fingerprint classification system. As its basis, he used the four basic fingerprint patterns defined by Galton: arch, whorl, and inner and outer loops. Unlike Galton, however, who used the internal and external loop distinction only on forefingers, Vucetich used it on all ten. This provided him with 1,048,576 primary classifications in which to file his ten-digit fingerprint sets, nearly ten times as many as Galton's system. Vucetich also devised detailed methods of subclassifying each of the primary classifications. Using these subclassifications, he could categorize his fingerprint cards into small groups that were easily searched.

Eventually, the Vucetich system would spread throughout much of South America, though his methods remained unheard of in England until long after fingerprints had been adopted by Scotland Yard. Credit for the fingerprint classification system adopted in England and Europe would go, not to Vucetich, but to an unknown English police chief who, at the time of the Rojas murder, was looking for methods to identify renegade members of the so-called "criminal tribes" who wandered the hills of Bengal, India.

. . .

Edward Henry's first dealings with the nomadic hill tribes of India came in 1882 soon after his promotion to joint magistrate and deputy collector. The elder women of the Magahiya Dom threw themselves at his mercy, begging for an end to their persecution by the British colonialists. The Doms, like other ancient aboriginal tribes, existed outside the Hindu system of social control, worshipping deities and fearing demons of their own. They wandered from place to place, earning their livings by weaving cloth and making music. But their nomadic existence was an infraction of the Criminal Tribes Act. The British hunted them down and threw their men in jail. The tribe's women had grown weary of this existence, and they looked to Henry for relief.

Born in 1850, Henry grew up in London, and studied law at University College. He passed the entrance examination for the Indian Civil Service in 1871, and in 1873, he sailed with his first wife through the Suez Canal to begin his twenty-seven years in India, where he settled in Calcutta, Bengal. At first, Henry served in Bengal's revenue and judicial departments, and in 1882, the year he met with the Dom, he was promoted to joint magistrate and deputy collector. Enforcement of the Criminal Tribes Act numbered among his magisterial duties.

The concept of the criminal tribe had evolved out of British misunderstanding of the caste system. Castes, the basis of India's ancient social hierarchy, each had its own customs restricting the occupations and dietary habits of its members. Priests, for example, came from the Brahmanic castes, while merchants, traders, and farmers came from the Vaisya castes.

Edward Henry

The British erroneously reasoned that the existence of criminals in India implied the existence of whole criminal castes. If a member of a little-known nomadic tribe, therefore, stole a cow, the British assumed his whole tribe were cattle thieves.

Speaking before the Indian Colonial Legislature in 1871, a jurist, James Fitzjames Stephens, described the criminal tribe as one whose "ancestors were criminals from time immemorial, who are themselves destined by the usages of caste to commit crime, and whose descendants will be offenders against the law, until the whole tribe is exterminated or accounted for." That year, the British passed the Criminal Tribes Act, calling for the "registration, surveillance, and control of certain criminal tribes."

The colonialists enforced the Act against hundreds of wandering tribes, not because they were demonstrated criminals, but because their nomadic existence made them difficult to govern and control. The Act restricted the movement of these tribes and demanded that they obtain passes to travel from place to place. Violators of the Act's provisions earned themselves prison sentences, even if they were tradesmen and shepherds whose occupations required them to travel.

It was the application of the Criminal Tribes Act to the Magahiya Dom that brought the tribe's women before Edward Henry in 1882. The Dom were fed up. They wanted an end to the constant clashes with British-controlled police. Henry brokered a deal with them. He would provide land if their tribe promised to settle on it, giving up its tradition of wandering. This way, Henry's police could easily monitor and control a tribe that they believed was constantly up to no good, and the tribe's men would no longer be terrorized as long as they stayed on the settlement and obeyed the Act's travel prohibition. The women agreed, and Henry devised a plan to turn the "criminal" nomads into settled agriculturists. Two Dom colonies were founded in 1883 and 1887.

After serving as joint magistrate and deputy collector, Henry went on to assist with famine arrangements in Tirhoot, acted as private secretary to the province's lieutenant-governor, and was promoted to magistrate and collector. But it was Henry's experience with the Dom that had taught him the difficulties of personal identification. Enforcing the Criminal Tribes Act required the identification of tribe members who had deserted their prescribed settlement. Henry became keenly aware that, other than the hit-or-miss use of personal recognition, no reliable method existed for making these identifications.

The identification problem became even more urgent in the late 1880s when colonial officials began to realize that Indian criminals did not come only from certain castes or tribes. Though for some years they maintained the restrictions on the so-called criminal tribes, they also recognized that habitual criminals in India, like everywhere else in the world, could come from any segment of society. The British wanted to impose the travel restrictions of the Criminal Tribe Acts on habitual criminals, but to enforce these restrictions, the British needed to keep tabs on them. How could they without a reliable way to identify them?

At the time this question was most troubling the colonial government, in 1891, Edward Henry at the age of forty-one was promoted to inspector-general of police of Bengal. Having pondered the identification dilemma since his confrontation with the Doms, Henry saw a solution: the French system that had lately earned so much publicity in Europe. In March 1892, the year after he became inspector-general of police, Henry introduced an adapted form of Bertillon's anthropometry throughout the Bengal province.

In the system's first year, Henry's men measured over 500 members of the "notorious Burwar tribe." Members of this "criminal tribe," according to Henry, regularly deserted their settlements, and "come to this province, where they commit crime, concealing their identity by assuming disguises and false names." Henry also initiated the measurement of convicted criminals, putting the system to more noble use. In the first year of operation, measurements identified 112 out of 536 otherwise unidentifiable prisoners. By the end of the second year, Bengal police fixed the identities of one in three unidentified prisoners by means of anthropometric measurements.

Measuring the Burwar Tribe

Meanwhile, back in England, William Herschel had sent a copy of Galton's *Finger Prints,* along with an enthusiastic letter, to his old colleague Henry Cotton, who was now the chief secretary to the government of Bengal. Herschel recommended testing the fingerprint idea in various departments of the Bengal government. Cotton forwarded Herschel's letter to Henry, whose staff, by this time, had become frustrated with certain aspects of the anthropometric system.

In the Indian adaptation, each criminal's measurement card was distinguished from other cards in its classification by notations of distinctive marks observed on the criminal's body. Sometimes these notations were not sufficiently accurate to identify a prisoner. A prominent birthmark on the elbow, say, might not be mentioned on the measurement card and, therefore, the card did not provide definitive proof of identity. The problem arose, Henry and his staff discovered, because the importance assigned to each physical mark depended on the person taking the measurements.

Henry needed a less subjective characteristic to add to his cards. Bertillon used photographs, but the cost of providing equipment and training to each of Bengal's many police stations made this option impossible. The suggestion contained in Herschel's letter to Cotton, however, fit the bill. Thumbprints were neither expensive, like photography, nor subjective, like distinctive marks. "When thumb impressions are compared by a person with some understanding of the system of patterns," Henry wrote in an 1894 report to Cotton, "doubt is at once cleared away."

Though two more years would pass before Henry's police began experimenting with the use of all ten fingerprints, the use of the thumbprints alone as a final confirmation represented a forward step. By April 1, 1894, Henry's police had recorded the measurements and thumbprints of 11,000 convicts. The system identified known criminals throughout the region, whether they traveled or not.

In one case Henry reported, a criminal, using a mild poison known as *dhatura,* drugged and robbed two victims in the Darbhanga district within a two-day period. A local detective determined that the robberies fit the profile of Runjit Dusadh, a "registered poisoner and ex-convict." Dusadh had absconded and could not be found. Weeks later, police removed some travelers from a train at Dacca, 500 miles away, who were apparently so helplessly drunk that they had to be hospitalized. In fact, doctors found, the intoxicated passengers were not drunk at all. They were victims of *dhatura* poisoning. Incoherent, the passengers could give no description of their poisoner, but luckily, the police found a stranger who had been seen in the company of the drugged men and arrested him. He gave his name as Singh.

The local police took his anthropometric measurements, impressed his thumbprint, and sent the details to the central office. A day later, word came back that the suspect had been identified as Dusadh, the man wanted in the Darbhanga poisoning cases. The Indian anthropometric system, adapted to include thumbprints, had snagged Dusadh, and he was convicted of both crimes. The irony was that, while thumbprints definitively identified the likes of Dusadh in the backwaters of India, back in the seat of the Empire, London police officers still fumbled with the inaccuracies of personal recognition. And the system was falling apart.

. . . .

Early on the morning of the fourth of May, 1893, a London constable found Percy Albert Blake rummaging through a pawnbroker's shop on the Strand as though he intended to burgle it. When questioned, Blake gave an incoherent and rambling explanation of his presence in the shop. He was either a lunatic or a very clever burglar pretending to be one. The constable arrested him and investigated further. If Blake was a career criminal, his likeness would be in the Yard's photographic register of known criminals.

One photograph in the register, the constable thought, looked exactly like Blake. It bore the name of Henry Steed, who, in 1881, had been convicted of attempted burglary and sentenced to Pentonville Prison. Steed's arresting officer and a Pentonville jailer both confirmed that Blake was the same man. Though neither of these men had seen Steed for over ten years, the police gave their identifications more credence than Blake's protestations. He had lied about his identity, the police decided, and his confusion in the pawnshop was just an act. Henry Steed's file showed that his leg had once been broken. A doctor testified that neither of Blake's legs showed any sign of fracture. Still, the police did not relent.

They made a terrible mistake. Blake was just what he appeared to be: a lunatic who, in his delusional state, had wandered through the unlocked door of the pawnshop. The police, however, prosecuted him for burglary. To make matters worse, if convicted under the name of Steed, Blake would be labeled a habitual offender and would receive the harshest penalty the law could muster. Blake sat in jail for several more weeks before his defense finally proved the truth of his identity and his behavior.

Blake's narrow escape from false conviction numbered just one among many of the period's cases of mistaken identity. In 1889, police mistook David Callan, a beggar, for someone who had eleven previous convictions. They charged him with being an "incorrigible rogue." In the same year, a jailer erroneously testified that James Coyle, convicted of larceny, had served time in Millbank. Coyle was wrongly sentenced as a repeat offender. In 1891, police misidentified Eliza Witchurch, a burglar, and credited her with another woman's previous convictions. Witchurch undeservedly received an extra seven years on her sentence.

Britain's ability to identify its citizens had not improved one bit since the days of the Tichborne Claimant. The terse physical descriptions in England's first annual register of habitual criminals, published in 1869, were basically useless. The photographs added to the register after the Prevention of Crimes

Act of 1871 resulted, by the 1890s, in an overwhelming avalanche of over 115,000 pictures. A separate register of distinctive marks tried to distinguish criminals by their bodily peculiarities, but instead gave long lists of scofflaws who shared the same characteristics. It listed twenty-eight criminals, for example, who had rings tattooed around the second fingers of their left hands.

In the late 1880s, the police picked up no scent of Jack the Ripper, who savaged one London prostitute after another without getting caught. Only three years later, in 1891, the poisoner Thomas Neill Cream notched up six victims before the police investigation narrowed in on him. Criminals, it seemed to the public, could work their evil crafts among the throngs of London with no fear of recognition by the police.

Recognizing the problem, Sir Richard Webster, Q.C., a Member of Parliament, persuaded the Home Secretary, Herbert Henry Asquith, to consider adopting Bertillon's much-vaunted scientific method. Webster had visited Bertillon and had seen the French system's production-line identification of criminals with his own eyes. In the same period, the Inspector of Prisons, Edmund du Cane, proposed the adoption of fingerprints. A friend of Francis Galton's, du Cane personally presented Home Secretary Asquith with a copy of Galton's book *Finger Prints*.

The result was that on October 21, 1893, the Home Secretary appointed Charles Troup, a senior civil servant in the Home Office, to chair a committee to investigate the two identification options. In making its decision, the Troup Committee visited jails, operated the photographic and distinctive marks registers, and interviewed countless police officers and scientists. But first it sailed across the channel to watch Alphonse Bertillon demonstrate how he had recently identified one of France's most famous murderers.

A year earlier, in 1892, a violent explosion had rocked the

Boulevard Saint-Germain in Paris. As smoke poured from the shattered windows of number 136, police and firemen, looking for victims trapped under the rubble, discovered the remnants of a bomb. A terrorist group, the police surmised, had made an attempt on a state judge, one of the building's residents. Sixteen days later, another bomb exploded at 39 rue de Clichy.

The stairway demolished, inhabitants screamed for help from their shattered windows, a state prosecutor among them. The judge and the prosecutor had worked together in the recent trial of a group of anarchists. Obviously, the bombings were an attempt at revenge. Suspicion settled on an anarchist supporter known as Ravachol, but the police could not find him. Meanwhile, the anarchist newspapers called him a hero. Six days after the second bombing, a restaurant owner sent word to the police that one of his diners, who fit Ravachol's description, conversed passionately with a waiter about anarchism. Five policemen were waiting outside the restaurant when the man emerged. He drew a revolver and tried to escape, but the police overpowered him. Ravachol struggled against them all the way to the police station. "Follow me, brothers!" he cried. "Long live anarchy, long live dynamite!" The problem for the authorities would be dealing with Ravachol without martyring him to the anarchist cause.

The police dragged Ravachol, streaming with blood, to Bertillon's office, but his struggling made it impossible to take his measurements. The following day, Ravachol had calmed down and affected a hero's demeanor, obligingly allowing Bertillon to measure him—Height: 1.663 meters; spread of arms: 1.780; and so on. Bertillon discovered that Ravachol's measurements and description precisely matched those of Claudius François Koenigstein, a man wanted for killing three people during two separate burglaries. If Ravachol and Koenigstein were the same man, Ravachol was anything but a hero.

Left-wing newspapers protested indignantly at Bertillon's identification. The police, they said, wished to represent their hero as an ordinary criminal who killed for gain. But, accused of the bombings before a tribunal, Ravachol admitted he was Koenigstein. He also admitted to the robberies and murders. Thanks to Bertillon's identification, Ravachol's execution would make no martyr. "If you want to be happy," sang Ravachol on the way to the guillotine, "hang your masters and cut up the priests."

Bertillon regaled the visiting Troup Committee with this tale and others. At his office, he proudly guided them to the heart of his system, the huge wooden cabinet containing 243 drawers and about 40,000 anthropometric cards. The cabinet was divided into three compartments according to classification of head length: long, medium, or short. Each compartment then contained nine sections in three rows, arranged by broad, medium, and narrow head-width categories, and three columns, arranged by long, medium, and short left middle fingers. Each of the sections then contained nine drawers arranged in three rows, classified by the length of the left foot, and three columns, classified by the length of the forearm.

The cards were so carefully organized that, after taking their measurements, Bertillon's clerks went straight to the correct drawer, picked out a bundle, and scanned only seven cards on average before finding the one that matched their prisoner. If no card was found, the clerk could confidently conclude that the prisoner had never been measured before and was a first-time offender.

The accuracy of the system had been proved when Bertillon's Department of Judicial Identity was established in 1888. Bertillon's superiors had insisted that his assistants pay a ten-franc reward to any prison warder who recognized unidentified old offenders. In 1889, 31,000 criminals passed through Bertillon's department, but only four warders earned a reward.

In 1890, there were also four failures, the next year, six, but in 1892, no misidentifications were detected. By the time of the Troup Committee's visit, the certainty of being identified had caused most criminals to give up using false names. Only one criminal in fifteen even tried to fake his identity. France had found a surefire way to identify its habitual criminals.

Until Bertillon put his precious measuring calipers in the Troup Committee members' own hands, his system utterly impressed them. But when they tried to work the system themselves, they were stumped. Over and over, Bertillon's clerks had to demonstrate how to place the caliper tips on a prisoner's head to ensure measurement consistency. What seemed like hundreds of little nuances frustrated the Committee as they tried to measure the prisoners. They could not measure one prisoner to completion without being reminded of something they'd missed or done wrong. Weeks of training would be required for the British officers who might run this system, the Troup Committee worried.

On the other hand, back in England, Troup and his colleagues were fascinated by the ease of taking fingerprints. They looked on as Francis Galton trained a Pentonville Prison warder in the technique. In almost no time, the warder had mastered the system, and he took thirty-five sets of impressions from prisoners in under an hour. Fingerprints had the advantage, since they were taken directly from the body, of being impervious to mistake. In their final report, Troup's Committee wrote that "We are clearly of the opinion that for the purpose of proving identity the finger prints examined and compared by an expert furnish a method more certain than any other. They are incomparably more certain than personal recognition or identification by photograph."

But they were disappointed in Galton's unsophisticated system of classifying fingerprints. Galton had not refined it suffi-

Taking a head measurement

ciently for practical use in a large register. Fingerprint sets tended
to lump themselves into a very small number of his system's categories. In Galton's own collection of prints, for example, 12
percent of the cards fell into only 1 percent of the classifications.
The most common classification, the one pertaining to all loops
on all the fingers, contained 6 percent of his total collection. In a
criminal register containing, say, 90,000 cards, this would mean
that a single classification would contain as many as 5,400
cards, far too many for a policeman to easily scan through.

The solution, the Troup Committee decided, was to combine
the best aspects of both Bertillon's and Galton's systems. Britain
should adopt an identification system based on the combination
of Bertillon's classification system and Galton's absolute identification by fingerprints. Bertillon's cabinet of 243 drawers would
be adapted along with his five major measurements, but his six
finer measurements would be rejected. Cards would be assigned
to the various drawers according to their classification by measurements, and within each Bertillon drawer, they would be subclassified by fingerprints according to Galton's system. Just as in
Henry's system in India, comparison of the prints themselves
would provide the last word on a prisoner's identification. The
Troup scheme, however, was more advanced than Henry's since
it used not only thumbprints but ten-digit sets.

On October 4, 1894, eight years after Henry Faulds first
argued the merits of fingerprints with the junior detectives of
Scotland Yard, and fourteen years since he published his *Nature*
paper, the Troup Committee published their report suggesting
an adaptation of his system. Fingerprints, the report said, were
"now associated with the name of Francis Galton" but were
"first suggested and to some extent applied practically by Sir
William Herschel." Dr. Henry Faulds, who for so long had campaigned for the adoption of fingerprints, was not mentioned on
any of the report's sixty-seven densely printed pages.

Eight

The Case of the
Little Blue Notebook

Henry Faulds quickly tracked down Charles Troup in the pala-
tial Government office buildings of Whitehall. The Troup Committee had made a terrible mistake, Faulds thought. They had
wrongly concluded that fingerprints needed the extra classification muscle of Bertillon's measurements to carry the weight of a
full-scale criminal register. But the cumbersome measurements
could be entirely disposed of, Faulds believed, if only the police
used his more sophisticated method of fingerprint classification.
For this reason, Faulds had rushed by train to London from his
new home near Birmingham.

By the time of the Troup Committee, Faulds had long
accepted defeat in his quest to have fingerprints officially
adopted. When his letters to police chiefs around the world, his
visits to Scotland Yard, and his interview with Chief Inspector
Tunbridge had all led nowhere, Faulds had no choice but to
return to the more pressing business of supporting his family.
He moved with them to Fenton in Staffordshire, near Birmingham, where he took over the practice of a "club doctor," tend-
ing to poor people's illnesses in return for the penny or so
subscriptions they paid each week. As official surgeon for the

Fenton police, Faulds also provided expert medical testimony in criminal cases at the local police court.

Living in this Midlands backwater, Faulds probably first stumbled over the Troup Committee's existence in a newspaper report. He must have celebrated their proposed use of fingerprints along with measurements, assuming, at first, that his persistent pestering of Scotland Yard officers had finally reaped reward. Having obtained his own copy of the Troup Committee report, he was deeply disappointed to find no mention of either his name or the fact of his original suggestion of the system.

Troup and his colleagues had interviewed thirty separate witnesses during their investigations, but Faulds did not count among them. The result was the adoption of Galton's insubstantial system of fingerprint classification that, as Faulds would later write, did nothing more than bunch fingerprints "up in similar bundles, like chocolate cakes." Faulds had tirelessly campaigned for fingerprints for fourteen years, and yet he had been invited neither to describe his more sophisticated method of classification nor to discuss any other aspect of the system he considered himself to have invented. He felt deeply excluded.

As he sat in the overstuffed chair in Troup's office, Faulds must have strained to contain his disappointment and hurt at the oversight. He tried to explain why his own classification system, not Galton's, should be adopted. Faulds's system grouped each fingerprint into twenty-one categories, he explained, compared with Galton's only three. Applied to ten-digit fingerprint sets, these categories resulted in nearly 17 trillion classifications, while Galton's system provided only about 100,000. Faulds's system, he tried to convince Troup, provided such great differentiation between fingerprint sets that Bertillon's measurements would not be required. Perhaps because Galton had already undermined his reputation, Faulds's pleading came to nothing.

He asked Troup why the Committee reported that William

Herschel was the first to suggest fingerprints. Why had they not recognized that it was Faulds, not Herschel, who first published the fingerprint idea in *Nature*? Why hadn't they mentioned Faulds's numerous meetings at Scotland Yard? Troup had never been told about Faulds's meetings with Chief Inspector Tunbridge, Troup told Faulds, and, therefore, the Committee's fingerprint information had come entirely from Francis Galton's testimony and book.

Faulds had never, until now, heard of Galton's *Finger Prints*. He immediately went in search of a copy, and urgently scanned through its pages. He was highly impressed by the level to which Galton had raised the fingerprint science. He felt pleased that, by patient experiment, Galton had carefully examined Faulds's suggested use of fingerprints both in ethnology and in the field of criminal identification. "You must always deserve the greatest credit for your lucid and masterly exposition of the subject," Faulds wrote in a letter to Galton dated December, 1894.

What troubled Faulds about the book was the minor position Galton assigned to him in a list of fingerprint pioneers. Galton gave him virtually no credit. In Chapter Two, "Previous Uses of Finger Prints," Galton applauded the well-known engraver Thomas Bewick for being "the first well-known person who appears to have studied the lineations of the ridges as a means of identification." In fact, Bewick had merely adorned the frontispieces of his 1804 and 1818 ornithology texts with woodcuts of his thumb impression. He did not study the use of fingerprints as a means of identification at all.

Galton then glowingly described two extremely limited American applications of the fingerprint conception. In the first, geologist Gilbert Thompson, the head of a geological survey in New Mexico, used his thumbprint in 1882 as a seal to forestall forgery of his signature on payroll money orders. In the second, in 1888, California photographer I. W. Tabor proposed to Congress the use of thumbprints to register Chinese immigrants. His suggestion came to nothing.

Despite Galton's enthusiastic reports of the Americans' work, they had made no contribution to fingerprint science. Faulds's study was far more substantial, and it had come earlier. Galton, however, barely tipped his hat at Faulds's work, saying only that "occasional instances of careful study may also be noted, such as that of Mr. Fauld [sic]." Faulds felt highly insulted that Galton had not bothered with the correct spelling of his name. He was even more horrified by the pride of place Galton had assigned to William Herschel.

Galton's book rambled on about Herschel's fingerprint work for more than a page. No fingerprint study was "comparable in importance to the regular and official employment made of finger prints by Sir William Herschel, during more than a quarter of a century in Bengal," it erroneously proclaimed. Herschel had, in fact, used fingerprints officially for less than two years. "Sir William Herschel must be regarded as the first who devised a feasible method for regular use," Galton insisted. While it was true that Herschel had used fingerprints as signatures, he had never, like Faulds, designed a classification system nor publicly suggested the fingerprint identification of unknown criminals. He could hardly be said to have "devised a feasible method for regular use."

Faulds felt his ideas had been stolen. Galton's history chapter was a transparent attempt to distance his work from its reliance on Faulds's original ideas. For the first time, the effects of the then secret 1888 pact between Herschel and Galton became painfully clear. Herschel would happily admit, in a 1909 letter to the *Times*, the existence of the pact: "When communicating to [Galton] in 1888 . . . I merely stipulated that he would recognize the fact of my putting the finger-print system into full and effective work . . . as early as 1877, after some 20 years' experimenting for this one definite purpose. He did more than keep his promise at all times . . . He assigned to me the priority of devising and adopting officially a feasible method of turning the finger-marks to practical use for identification." A sort of scientific kingmaker, Galton had crowned Herschel the inventor of fingerprints. In doing so, he divorced Faulds from the credit he deserved for his fourteen years' work.

Galton happily gave credit to Tabor and Thompson, because their uses of fingerprints were much less interesting than Galton's own. Faulds's ideas, however, competed with Galton's, and this is why, in Faulds's mind, Galton had chosen to belittle his work. What must have galled Faulds, in particular, was the fact that Galton had been privy to Faulds's private communication on the fingerprint subject—his letter to Darwin. Galton knew personally of Faulds's early work, yet still minimized Faulds's importance. A couple of high-society chums, Galton and Herschel had completely erased Faulds's presence from the fingerprint picture.

Faulds determined to set the record straight. He wrote a furious letter to the editor of *Nature,* published in October 1894 under the title "On the Identification of Habitual Criminals by Finger-Prints." In it, Faulds rightfully pointed out that "as priority of publication is generally held to count for something," and as his was "absolutely the first notice of the subject contained in English literature," he should retain some credit alongside Herschel and Galton. He also mentioned, fairly, that he alone had suggested in 1880 the registration of all ten digits, the approach adopted by the Troup Committee, while Herschel had variously used the impressions of only one or two digits.

Faulds's hurt and anger spilled out all over his letter. His most important points—his own priority and desire to be recognized for it—got lost in a barrage of insults and petty grievances against Galton and Herschel. He chastised Galton for misspelling his name in *Finger Prints*. He questioned Herschel's claim to have used fingerprints officially as signatures in India, barely refraining from calling him a liar. He called into question the existence of an 1877 document that Herschel claimed could prove his early use of fingerprints. "A copy of that semi-official report would go far to settle the question of priority," Faulds wrote.

Faulds had publicly challenged Galton and Herschel to a bare-fisted fight, not only over the origin of fingerprints, but over their honor as gentlemen. It was a mistake. Herschel rose to Faulds's challenge, and produced the 1877 document, a letter he had written to the Bengal Inspector-General of Gaols. In it, he described his use of fingerprints as signatures and suggested their widespread adoption. The "Hooghly Letter," as it came to be called after the town where Herschel wrote it, haunted Faulds for the rest of his life. Galton and others repeatedly used it to falsely contradict Faulds's claim to priority. The Hooghly Letter hadn't discussed either the registration of ten-digit fingerprint sets to identify criminals or the potential use of fingerprints as crime scene evidence. But its existence distracted from the fact that Faulds's *Nature* letter remained the first known written notice—published or not—of the fingerprint conception as it is used today.

Herschel first published the Hooghly Letter in a November 1894 edition of *Nature*. His correspondence also contained, in blistering terms, his answers to Faulds's assertions. "Mr. Faulds' letter of 1880 was, what he says it was, the first notice in the public papers . . . of the value of finger-prints for the purposes of identification . . . At the same time I scarcely think that such short experience as that justified his announcing that the finger-furrows were 'for-ever unchanging.' . . . The position into which the subject has now been lifted is therefore wholly due to Mr. Galton through his large development of the study." Reading between the lines, Herschel's meaning was that he was first, Galton was second, and Faulds was irrelevant.

The truth was that both Faulds and Herschel had made valuable contributions. Faulds had a broader vision for fingerprint use, while Herschel, by using them officially, had given the idea credibility. Herschel's forty-year-old fingerprint collection proved fingerprints were persistent with age. Faulds, by removing the ridges in various ways and observing their regrowth, had proved that fingerprints withstood injury. Taken together, their results proved the complete reliability of fingerprints.

For now, however, Faulds's contribution remained ignored and disputed. Because Faulds's temper had got the best of him, Herschel had won the first volley in their pen duel, though it would continue on and off for the next twenty-five years. But while Herschel and Faulds squabbled over their relevant importance, one of Edward Henry's staff members, in India, had begun the process of developing their ideas along with Galton's to make fingerprints practical for application to police work.

· · ·

Ever since Edward Henry had ordered the addition of thumbprints to Bengal's anthropometric cards, Azizul Haque, head of Henry's identification department, had hoped to use fingerprints exclusively to identify criminals. Haque had never liked the anthropometric system that Henry had introduced. Haque had met frustration after frustration administering it. Levels of literacy and arithmetic ability were low in the region's police stations. Even after repeated training sessions, Bengal's police officers could not maintain sufficiently high standards of measurement accuracy to run the system efficiently. Haque, who was an advanced mathematician, saw the potential elegance in Bertillon's system, but trying to make it work was like trying to play a musical masterpiece on an out-of-tune piano.

Born in Paigramkasba, Bengal in 1853, Haque studied math and science at Presidency College in Calcutta. In 1892, Edward Henry wrote to the college principal asking for the recommendation of a strong statistics student, and the principal nominated Haque. Henry recruited Haque as a police sub-inspector and, initially, gave him responsibility for instituting the anthropometric system in Bengal. Haque performed the statistical studies necessary to apply the anthropometric system's small, medium, and large classifications to the Bengalese population. Even his fastidious work could not overcome the police force's lack of numerical skill.

At Haque's central office, clerks under his command had to assume that the measurements they received from around the region contained errors. They could not be certain, therefore, that a criminal's card was filed in its correct category. They had to rummage through many pigeonholes during every search, negating the very purpose of the system. Taking a set of measurements or searching for a prisoner's card each took an hour, many times longer than the process required in Bertillon's office in Paris. Haque was not satisfied. With Henry's permission, in the hope of eventually replacing measurements, Haque began experimenting with ten-digit fingerprint sets.

During the same period, in the months following the release of the 1894 Troup Committee report back in England, Haque's enthusiasm over fingerprints encouraged Henry to strike up a correspondence with Galton, who busily worked to refine his classification system in the hope of its more general adoption. In October 1894, while in England on home leave, Henry visited Galton at his anthropometric laboratory, where they discussed the advantages fingerprints would offer if a workable system of classification could be found. Galton and Henry hatched an agreement to collaborate on the advancement of fingerprints. Galton would keep Henry abreast of refinements he made in his fingerprint classification system; Henry would supply Galton with sets of convict fingerprints for his experiments and attempt to apply Galton's ideas in a practical setting.

At first, Haque was overjoyed, and, looking forward to the practical system of classification Galton had promised, he enthusiastically sent him fingerprint sets. In 1895, Galton published his book, *Finger Print Directories,* containing his improved fingerprint classification system. Haque read it eagerly, hoping to get under way with the exclusive use of fingerprints throughout Bengal. Galton's major innovation, Haque discovered, was the subclassification of loops and whorls. Loops were subclassified by "ridge counting"—enumerating the number of ridges between the center and outer boundary of the fingerprint core. Whorls were subclassified by categorizing them into twenty-eight groups according to peculiarities in their patterns.

Haque was discouraged. How could his barely educated police officers be expected to run such an intricate system? Counting loop ridges in a near-microscopic image was rife with potential error. Asking an identification clerk to distribute the

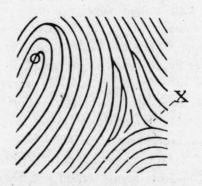

Counting the number of ridges in the core
(between the x and the o)

whorl patterns into twenty-eight separate categories, an extremely subjective process, was like asking a child to alphabetize the Greek dictionary. Galton seemed not to realize that a reliable system had to be simple enough to be run by common clerks, not by highly trained scientists. Haque, in his frustration, began work on a classification system of his own, borrowing elements from Galton's.

By 1896, two years after the Troup Committee had made its report in England, with some help from Henry, Haque had settled on a strategy that showed promise. Henry, his boss, had enough confidence in Haque's progress to order that criminals' ten-digit fingerprint sets should be taken along with their anthropometric measurements. Haque and his staff taught police clerks around the region to take criminals' fingerprints on a specially designed slip, each digit printed in its natural order from thumb to little finger, with the fingerprints of the right in a line above the fingers of the left hand. Fingerprint slips poured in to Haque's office, and they contained far fewer errors than the anthropometric cards. Haque was overjoyed.

Haque filed his slips in a large wooden cabinet he had built especially for his fingerprint classification experiment. Haque's cabinet, containing 1,024 pigeonholes in thirty-two columns and thirty-two rows, was the heart of his system. Its use required no math and no measurements. First, Haque labeled each of a criminal's ten fingerprints as either a loop (L) or a whorl (W), a simple distinction that any layman could make. Haque also lumped the arches previously contained in Galton's Arch-Loop-Whorl system in with the loops. He then wrote out the pattern labels in pairs, for example: WL-LW-LW-LL-WL.

Haque segmented his cabinet into quadrants, each corresponding to the arrangement of patterns on the first fingerprint pair. Referring to the figure below, LL, in the first pair of fingerprints, corresponded to the upper left quadrant of the cabinet,

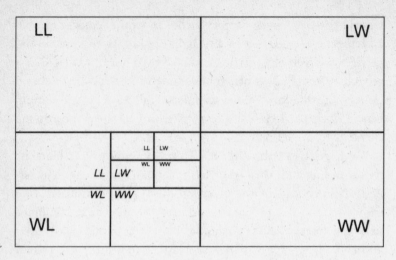

Segmentation of Haque's cabinet

LW to the upper right, WL to the lower left and WW to the lower right. Haque similarly quartered each quadrant according to the patterns on the second pair of fingerprints. He repeated this successive quartering of the pigeonholes into sections for each of the third, fourth, and fifth pairs. By this system, the five fingerprint pairs in a set would lead directly to an individual pigeonhole. The fingerprint set labeled WL-LW-LW-LL-WL, for example, ended up in the pigeonhole in the thirteenth column and the eighteenth row, which could be represented by the fingerprint set's "classification formula" as 13/18. Haque also devised an equation for determining a fingerprint set's classification arithmetically without referring to the cabinet.

By 1897, Haque had collected 7,000 fingerprint sets in his cabinet. His simple methods of further subclassification, which were easier to learn and less prone to error than Galton's, meant that even a collection numbering in the hundreds of thousands could be divided into small groups of slips. As he predicted, his

Right thumb.	Right index.	Right middle.	Right ring.	Right little.
(Fold) W	\	\	W	\ (Fold)

LEFT HAND.

Left thumb.	Left index.	Left middle.	Left ring.	Left little.
W	/	/	W	/

A fingerprint set (slashes are loops)

fingerprint sets, compared with anthropometric cards, were far less prone to error and could be classified and searched with much greater confidence. The registration of a convict or a search for his existing card took an hour under the anthropometric system, but only five minutes using Haque's classification of fingerprints.

Haque's boss, Edward Henry, was overjoyed with Haque's results, and Henry saw that they would reflect well on him and his career. He asked the colonial government to convene a committee to evaluate the system for widespread use. The committee reported that fingerprints were superior to anthropometry

"1. In simplicity of working; 2. In the cost of apparatus; 3. In the fact that all skilled work is transferred to a central or classification office; 4. In the rapidity with which the process can be worked; and 5. In the certainty of results." Fingerprints, in other words, were the new hero in criminal identification.

In June, 1897, the Governor-General directed that the identification of criminals by finger impressions should be adopted throughout India, the first country in the British Empire to adopt the fingerprint system of identification first suggested by Faulds and Herschel. The 150,000 to 200,000 anthropometric cards collected in British India's various provinces were each replaced with fingerprint slips as rapidly as possible. Because India's population, at 200 million, was so much greater than England's, it was a more than sufficient proving ground for the system's eventual adoption by Scotland Yard. Within a year, it had caught its first murderer.

. . .

On the morning of August 16, 1897, Hriday Nath Gosh, a manager of a tea garden in the Jalpaiguri district of northern Bengal, was found murdered in the bedroom of his bungalow. Gosh's personal effects were scattered around the room, and his iron safe had been robbed of several hundred rupees. Among the disorder, police found a kukri knife, a sort of dagger common in Nepal, with a short, curved blade. The murderer had used it to slit Gosh's throat.

There were no witnesses to the crime, and the only evidence found by the police was a little blue Bengali notebook. On its cover were two clearly visible fingerprints in what appeared to be dried blood. Police also quickly heard rumors that a number of present and former employees of Gosh hated him for being a bully and a hard taskmaster. After interviewing his employees, the police settled on three main suspects: Gosh's cook, who had

The bloody print

been seen wearing a bloody apron on the morning of the murder; the family of a woman with whom Gosh, as Henry put it,
"had a liaison;" and a wandering gang of criminals who were
camped in the neighborhood.

Police had also suspected an ex-servant named Ranjan Singh,
also known as Kangali Charan, whom Gosh had prosecuted for
the theft of some money, landing Charan in Jalpaiguri Prison.
Charan had recently been released under a general amnesty on
the Diamond Jubilee of Queen Victoria, but had not been seen
in Jalpaiguri since. Police investigations showed he had returned
to his native town of Birbhum, far away in western Bengal, and
Charan was dropped as a suspect.

The blood on the cook's apron turned out to have come
from a pigeon he had slaughtered for the tea garden manager's
dinner; the criminal gang had been nowhere in the vicinity at the
time of the murder; and the police could find no evidence

against the family of Gosh's "liaison." Now, their only lead was the bloody marks on the little blue notebook that they sent to the Central Office of the Bengal Police in Calcutta along with their list of suspects. Though police had dismissed their suspicion of him, Kangali Charan's right thumbprint matched the marks on the book.

Charan was arrested in his Birbhum home, hundreds of miles from the scene of the murder, and taken to Jalpaiguri to stand trial. Witnesses, it turned out, had seen him at fairs and markets in the vicinity of the tea garden in the weeks before the murder. Charan's prison mates had heard him threaten his revenge on Gosh. He also had more money at the time of his arrest than he could account for in earnings. In May 1898, Charan was indicted for murder and burglary, and he pled not guilty, triggering the first known use of fingerprints as evidence in a trial.

The inked print taken in court

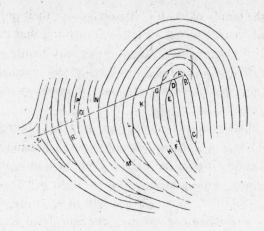

The points where the prints matched

In front of the judge and two "assessors," prosecutors took the inked impression of Charan's thumb and showed them point by point how it matched the impression on the notebook found in the murdered man's house. But the judge and assessors were reluctant to be the first to condemn a man to death on the basis of this new form of evidence. They argued that, while the thumbprint might prove Charan's presence at the murder scene after Gosh's death, it did not prove that it was Charan who actually slit his throat.

Furthermore, the judges and assessors argued, the murder weapon, the Nepalese knife, had been wielded skillfully, as though by a native of Nepal, which Charan was not. This, they said, raised doubts in their minds as to whether Charan was the murderer. Therefore, they found him guilty only of burglary, and he was sentenced to two years. It was a small victory, at least, for the fingerprint system.

Meanwhile, Henry's Bengal police department identified habitual offenders by fingerprints hand over fist. The largest number of identifications by anthropometry in a single year had been 334 in 1896, the fourth full year of the system's operation in Bengal. The fingerprint system outstripped that number in only its *first* full year of operation with 345 identifications in 1898. Its success grew still further in 1899 with 569 repeat offenders identified in Bengal by fingerprints.

Meanwhile, Henry had begun to tell those who asked that it was he who had come up with the classification system in a sudden flash of inspiration on a train, when he had no paper and had to resort to noting his ideas on his shirt cuff. The tales got back to England, along with word of the success Henry achieved on the backs of Haque and Galton. The news helped convince British authorities to follow the Indian example of adopting Faulds's and Herschel's scheme.

Nine

An Innocent in Jail

Back in England, Dr. John Garson, the scientist charged with running the new system recommended by Troup, scored his first sensational success. In a case involving the theft of a clock from a pub, police had collared a waiter who said his name was James Coots. A prison warder, however, recognized him as a seasoned crook called Frank Foley, who had been to prison three times. Coots denied the identification. Should the judge believe Coots, or the warder? In the past, his decision would have amounted to a flip of a coin, but this was just the sort of identity mystery Garson could now solve.

Garson was vice president of the Anthropological Institute, and England's foremost expert on anthropometric measurements. Since July 1895, when Scotland Yard hired him as scientific adviser, he raced around the country, hastily training warders and constables in measuring bodies and impressing fingerprints. Back in London, where all the identification cards were sent, he trained three police detectives in the methods of classification and search. Their new department came to be known as the "Metric Office," after the metric system by which the measurements were taken.

By the time of Coots's contested recognition in 1898, the Metric Office had made many routine identifications of repeat criminals. The Coots case, however, was the first in which Garson settled an identification dispute in court. He produced measurements and fingerprints of "Frank Foley," compared them with Coots's, and showed that they matched. Coots got three years, and Garson won his first public victory.

Still, the Metric Office had its detractors. One of them, Melville Macnaghten, the CID chief constable, complained that Garson's successes came too few and far between. Garson had failed to win over the police force's old guard to the advantages of his complicated scheme. They thought the measurements were often too much effort for too little reward, and they only used them in the most serious cases. As a result, Garson's department received only a trickle of search requests—822 in the whole year of 1900. Even when police troubled themselves to measure their scofflaws, the Metric Office, crippled by the small number of cards in its collection, managed to provide positive identification in only about one third of the cases they were sent.

As a result, police stuck to their old standby, personal recognition, and Macnaghten was horrified. He had served on the Troup Committee, and remembered cases of terrible suffering caused by the mistaken memories of warders and policemen. To any senior police or Home Office official who would listen, Macnaghten adamantly protested the continued reliance on personal recognition. Little did he realize that the wrongful recognition of Adolf Beck, a man who rotted in jail for the crimes of his look-alike, would later be one of the most disturbing cases in Macnaghten's career. Meanwhile, Macnaghten and his colleagues got wind of Edward Henry's successes with fingerprint classification in India.

Francis Galton arranged for Edward Henry, while he was on home leave from India, to present a paper at the 1899 meeting of the British Association for the Advancement of Science in

Dover. Galton, still believing Henry used the revised classification system Galton described in *Fingerprint Directories,* expected to share in the glory of Henry's success. Henry's triumphant record in India, Galton hoped, would help him promote his revised system to the government. But Henry did not discuss Galton's system at all. Instead, Henry extolled the elegant and superior classification methods devised in Bengal.

Galton had taken Faulds's ideas, Haque took some of Galton's, and now, Henry, describing the new classification system as if it were his own, took Haque's. By the time of the British Association meeting, Henry and Galton were like the last players left in a grown-up game of musical chairs. Henry gave his talk, the music stopped, and Galton was left standing. Henry had won. Galton, growing more and more agitated during Henry's presentation, had finally got a dose of the medicine he had earlier so roughly dosed out to Faulds. The methods of fingerprint classification, though Faulds, Galton, and Haque had each contributed to them, would forever be associated with Henry's name.

Within a year of Henry's presentation, Britain's Home Secretary convened a new committee to consider whether Britain should adopt the system that had nabbed so many habituals in India. However, the country's adoption of fingerprints was far from a foregone conclusion. The proceedings of the committee, chaired by Lord Henry Belper, became a battle between the two scientists who first established scientific identification in Britain, and the policeman who had finally perfected it.

The scientific naysayers, like jealous parents, would not let go of their grown-up child. Bitter at having been elbowed out of the limelight, Galton, in an article in the journal *Nineteenth Century,* implied that Henry had bent the truth of his results with fingerprints. "[It] seems to me so surprising that I should

greatly like to witness his methods tested on a really large col-
lection, say of 100,000," Galton wrote. In reply to a Belper
Committee question about how Henry subclassified whorls,
Galton replied bluntly, "I think it is a mistake."

The next to sneer at Henry's system before the Belper Committee was Dr. Garson of the Metric Office. Worried about
keeping his job, he objected in every conceivable way to the
independent use of fingerprints. More than a third of fingerprints taken by police were defective, he said. Identification staff
would suffer eyestrain from looking at the minute lines all day.
Police clerks did not have the scientific expertise to make positive identifications. His testimony, along with Galton's, shook
the Belper Committee's confidence in Henry's scheme.

Henry dissipated their worries when he rolled in a trunk
containing a set of 7,000 fingerprint slips. Over and over again,
before the Belper Committee's eyes, he quickly classified print
sets and made sample searches. He showed them the proofs of
his new book, *Classification and Uses of Finger Prints,* of which
the Indian Government had already ordered 15,000 copies. He
regaled them with the dramatic tale of the Tea Garden murder
case and the thumbprint that solved it. Most importantly, he
disproved the two main concerns raised by Galton and Garson,
establishing that there was no difficulty taking legible prints and
that a collection of 100,000 could be classified with precision.

The Belper Committee also quizzed a number of other witnesses, but in the end, it was the policeman's words against the
scientists'. The Belper Committee's members were Home Office
officials and magistrates who had experienced the difficulties of
Garson's system and the problems with personal recognition.
They knew something better was needed. Cautiously, the Belper
Committee came down on the side of fingerprints. They recommended establishing the scheme alongside the existing methods

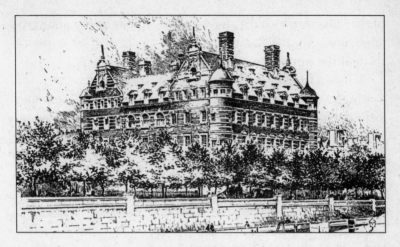

New Scotland Yard

until it could be determined "how far Mr. Henry's system could be safely adopted."

Triumphant, Edward Henry, appointed to the post of Assistant Commissioner of Scotland Yard, returned to Britain in 1901, both to establish the Fingerprint Branch and to take charge of the Yard's existing detective department, the CID. But just because Henry's system had won its adoption did not mean that it would be generally accepted by the public, the police, or the judiciary. It didn't help that when Henry quickly fired John Garson, the main opponent of his system, he unwittingly created a formidable enemy.

Garson, an acknowledged expert in the field of criminal identification, publicly argued that police clerks were not sufficiently educated to administer a system of scientific identification. His concerns gained him an ally in Henry Faulds, who thought the use of scientists would help avoid possible misidentifications. Edward Henry also faced the formidable opposition of the police force's old guard who had happily stopped the pre-vious identification fad in its tracks. Henry had to convince them of the worthiness of fingerprints.

What he would need, in the long run, was a big case, perhaps a sensational murder in which desperate, hardened criminals were publicly brought to justice by a fingerprint identification. For now, Henry took baby steps. He needed to rack up a few smaller victories. He won his first among the bustling crowds of a popular horse race.

. . .

Every June since 1780, the report of the Derby's starting pistol had echoed through the town of Epsom, Surrey, just west of London. Named for the twelfth Earl of Derby, the race was a fixture in the lives of the country's traveling pickpockets. The spectators made easing pickings for the light-fingered thieves.

The pocket-thief unlucky enough to get caught, having trav-eled to the Derby from afar, had the advantage of being unknown to the local police. He could swear in court that temp-tation bested him just this once, and get away with a slap on the wrist. All the thief had to do was lie about his name so police could not find any records to contradict his fictional story of innocence. This way, every year at the Epsom Derby, tens of pickpockets got away with lesser sentences than they deserved. Edward Henry intended to close that hole in the judicial net.

Within a month of his appointment as assistant commi-sioner, Henry had selected three men from the old Metric Office to kick off the Fingerprint Branch. He trained them, and the men then traveled the country giving instruction in the use of fingerprint forms. Though many of the nation's police consta-bles and prison warders viewed the system with caution, those who were open to it enjoyed its ease of use compared to the awkward and time-consuming taking of measurements. By the standards of Garson's Metric Office, the fingerprint slips and search requests poured in.

The Fingerprint Branch identified ninety-three repeat offenders in its first six months of operation. Henry's boss, the Police Commissioner, was so impressed that, before long, he canceled the viewing of prisoners on remand at Holloway. Macnaghten, who had worked under Edward Henry since he took over the CID a year earlier, was so overjoyed by the diminishing use of personal recognition that he would eventually dedicate his book to Henry. It was Macnaghten who suggested that the use of fingerprints at the Derby might help win the public's confidence in the system.

Two fingerprint men set up shop at the race, and took the prints of the fifty-four pickpockets arrested during the course of the day. They rushed the impressions to the Fingerprint Branch, where two officers beavered away through the night, searching through the collection for identifications. Twenty-nine of the day's collars, they found, were old offenders. A pile of records and photographs documenting the miscreants' past crimes greeted them as they filed into court the next morning.

The first defendant claimed his name was Green, from Gloucester, and that he had never been in trouble before. Up jumped one of Macnaghten's chief inspectors in protest. Fingerprints showed that the man in the dock was, in fact, Benjamin Brown of Birmingham, who had ten convictions to his discredit. "Bless the finger-prints," said Brown, "I knew they'd do me in!" Brown, like the other twenty-eight repeat offenders prosecuted that day, got double the sentence he would have received without his identification by fingerprints. The Fingerprint Branch had scored its first big victory, and later that month, it scored its second.

Investigating the burglary of a house in Denmark Hill, London, the Fingerprint Branch's Detective Sergeant Charles Collins discovered a complete set of fingerprints on the freshly painted sill of the window through which the burglar had come. The clearest impression, a loop pattern which Collins photographed, came from a left thumb. Because he had only one print, Collins could not sufficiently narrow down a search of the fingerprint collection, which was indexed by classifications based on all ten prints. Therefore, he asked the local police to come up with a list of burglars known to operate in the Denmark Hill area. He pulled the corresponding fingerprint sets out of his cabinet.

The matching thumbprint was in the file of Harry Jackson. When police tracked him down and questioned him, Jackson said simply, "I know nothing about the case." Sergeant Collins, now comparing a print taken directly from Jackson's left thumb with the one from the windowsill, knew differently. He had his man, and this was Edward Henry's first chance to take a burglar to an English court on the sole evidence of a fingerprint. Henry knew it was a risk, since most jury members would never have heard of fingerprints, but it was also another ideal opportunity to win public approval for the new system.

Knowing what was at stake, Henry, who had since been appointed Commissioner of the entire Metropolitan Police, wielded his influence to persuade the greatest prosecutor of his time, Richard Muir, to take the case. Eight years later, Muir would prosecute the difficult case against Dr. Hawley Harvey Crippen, who poisoned his wife and whose trial was considered the most sensational in Britain for fifty years. He would not normally have entertained a small burglary case like Jackson's, but Henry convinced him of its importance to the future of fingerprinting.

After spending many hours with Collins studying the intricacies of the fingerprint system, Muir opened the case against Jackson like any other burglary case, by explaining its circumstances. The case was an ordinary one, Muir said to the jury, except that "part of the evidence was of a kind which had never been used before a jury in an English criminal court." Muir quickly put Collins on the witness stand, and skillfully led him through a fascinating explanation of fingerprints. Collins told the jury how the system had first nabbed countless criminals in India, and how it was now in large-scale use in England. By the time he finished, the spellbound jury hung on his every word.

Only then did Muir ask Collins to exhibit the enlarged photographs of the thumbprint he found on the windowsill in the house in Denmark Hill. The burglar had left behind a mark from which Collins could tell the burglar's identity as though he'd left behind a calling card, Muir said. Next, Collins exhibited the enlarged thumbprint taken in ink directly from Jackson's left thumb. Collins pointed out each one of ten ridge minutiae that corresponded in the two diagrams.

"Can you say for certain to whom the print from the murder scene belongs?" Muir asked.

"The print on the window sash is that of the prisoner's left thumb," Collins concluded decisively.

"In your experience, have you ever found two persons having identical finger-prints?" asked Muir.

"Never," said Collins. Muir closed his case.

Harry Jackson insisted on his innocence. "I can conscientiously take my dying oath that I know nothing about the burglary at Denmark Hill," he told the jury, "and therefore the finger-prints could not have been mine." But looking at Collins's fingerprint evidence, the jury's own eyes told them, as though they'd seen a photograph, that the prosecution had its man. They found Jackson guilty, the judge gave him seven years, and the Fingerprint Branch won its second big success.

Wide publicity followed, with several newspapers carrying the case in detail, but overall, the result was not as good as Henry hoped. A letter signed by "A disgusted Magistrate" appeared in a London paper: "Scotland Yard, once known as the world's finest police organization, will be the laughing stock of Europe if it insists on trying to trace criminals by odd ridges on their skins." He referred to fingerprints as "half-baked theories some official happened to pick up in India."

Nor did it help the Fingerprint Branch when, in the same month that Jackson was convicted, September 1902, Detective Sergeant Collins appeared before another judge, this time to explain a mistaken identification. The Fingerprint Branch had erroneously advised police that the fingerprints of William Ward were identical with those of a criminal named Hopkins. It had been, Collins explained, not an error in identification, but in administration. A clerk had incorrectly copied down a reference number, causing the wrong file to be sent to the police. This explanation did little to settle public misgivings about the Yard's new scheme.

To make matters worse, Henry Faulds, the father of fingerprints, testifying that year before a committee convened to consider the fingerprint identification of army deserters, questioned the safety of convicting a criminal on the evidence of a single print. Since his 1894 hostile exchange with Herschel in the pages of *Nature,* Faulds had met with various Home Office officials, trying to garner some government acknowledgment for his role in the system's development. He'd gotten nowhere, fueling his bitter, vocal objections regarding the faults of the Fingerprint Branch.

Faulds told the War Department committee that, though he and others had compared many thousands of fingerprints sets to satisfy themselves that no ten fingerprints could be duplicated on two different people, no one had made a similar study to prove that each single fingerprint was unique. Though his motives were complicated, his point was well founded: Fingerprinting was a new science whose tenets had not all been proved.

Faulds argued that a single print register that filed, for example, right index fingers with right index fingers and left thumbs with left thumbs, needed to be established. Over time, this would allow the mutual comparison of single prints to prove that no two were the same. It would also allow single prints found at crime scenes to be quickly looked up. Until this was done, he told the War Department committee, there was a danger of sending innocent people to the gallows.

Though Henry and Macnaghten bitterly disputed Faulds's views, they still had not convinced many of their old-guard police officers that the new system was more than hocus-pocus. Personal recognition stubbornly kept its foot in the door, leaving a crack through which guilty men could slip. More importantly, it also left the door open for innocent men to get caught. That, much to Macnaghten's dismay, is what had happened in a nine-year-old case that suddenly came to his attention in 1904. The terrible case would prove to the country beyond any doubt how dangerous the lingering use of personal recognition could be.

• • •

At four in the afternoon on a cold winter day, fifty-five-year-old Adolf Beck, a wealthy Norwegian copper-mine owner, wearing a respectable-looking frock-coat and top-hat, stepped out of his door onto his front step on Victoria Street in London.

"I know you, sir!" a woman, Ottilie Meissonier, cried from the street below.

Beck descended the steps. "I beg your pardon, madam," he said, trying to push past the stranger. She barred his path. "What do you wish from me?" he asked.

"I want my two watches and the rings!" said Meissonier, a language teacher. The stout, mustached Beck had never before laid eyes on the woman. He sidestepped her and walked on. "I

Adolf Beck

will follow you wherever you go!" Meissonier shouted at his back, following him. Beck hurriedly headed off in search of a police constable to detain her.

It was December 16, 1895. More than a year had passed since the Troup Committee had recommended the adoption in Britain of the hybrid identification system based on both anthropometry and fingerprints, but Dr. John Garson was only beginning to collect his precious identification cards and it would still be some time before he made his first identification. Police still relied almost entirely on personal recognition, a fact that Adolf Beck would profoundly regret.

With Meissonier still at his heels, Beck at last found a policeman whom he hoped would detain her. Much to Beck's astonishment, Meissonier, in turn, accused Beck of having swindled her out of her jewelry. The constable arrested them both.

At the police station, Meissonier explained that a man matching Beck's description had pretended to mistake her for an acquaintance on the street three weeks earlier. They chatted for a while and discovered a common interest in flowers. When she told the man she had just received a box of chrysanthemums, he asked if he could come see them. "I gave permission for him to come next day," Meissonier said.

When this new acquaintance visited, he told Meissonier that he was a member of the landed gentry. Telling Meissonier that her language skills would be helpful, he invited her on a trip to the Riviera with a group of his friends. She would need a better wardrobe for the trip, he told her. He gave her a check for forty pounds to pay for some new dresses. He also asked to borrow her watch and rings for the purpose of sizing some better jewelry. He departed and Meissonier, overjoyed with her sudden good fortune, rushed to the bank to cash the check. At the cashier's counter, she discovered that it was worthless. She had been conned of her jewelry. She never saw the "lord" again, until three weeks later, when she "recognized" Adolf Beck coming down his front steps.

Beck was horrified at the allegations leveled against him. He vigorously denied them, but unfortunately for him, over the previous two years, twenty-two women had been defrauded in a similar way. The police had had no luck tracking down the con artist, but now, they believed, they had their man. If they were mistaken, they assured Beck, it would become clear when the other victims looked at him. Amazingly, Beck's luck turned worse. Ten of the women absolutely insisted that Beck was their perpetrator. "I would pick him out of a thousand," declared Fanny Nutt. "I am absolutely certain he is the man," said Alice Sinclair.

Meanwhile, the police became convinced that Adolf Beck's true identity was that of John Smith, a man who had in 1877 been arrested and convicted of robbing single women in the now familiar manner of impersonating a lord. Beck had not even been in England in 1877. He offered to bring witnesses from South America who would swear he was there. He did not know John Smith and had never spent a single moment in a British prison. Nor did Beck match Smith's description.

John Smith's file said he had brown eyes, while Beck's were blue. Beck's body carried none of the distinguishing marks listed in the file, either. Scotland Yard dismissed the inconsistencies as administrative errors. They had found the two retired police constables who had originally arrested Smith. Constable E. Spurrel, despite the passage of nineteen years, said: "The accused is the man. There is no doubt about it. I know what depends on my testimony and can say without a doubt that he is the man."

Beck protested, "I swear by my Creator: the women are mistaken, and the police are mistaken!" Even some of the witnesses expressed some doubt. Annie Townsend said, "When I hear him talk, I'm not so certain. In my home he spoke Yankee slang."

Ottilie Meissonier testified, "He had a scar on the right side of his neck," and admitted, when asked to point it out in court, "I don't see it now."

Beck's protestations and the women's misgivings did him little good at his Old Bailey trial before Judge Fulton. The parade of women identifying Beck from the witness box overwhelmed whatever doubts were expressed. The jury found Beck guilty after a three-day trial, and because the police insisted that Beck was John Smith, Judge Fulton sentenced Beck as a repeat offender to seven years, instead of the five years usually meted out to first-timers. Beck's life was in ruins.

At the prison, officials would not even allow him to be known by his own name. John Smith's prisoner number in 1877 had been D523, and that was the number assigned to Beck, amended by the letter W to indicate a repeat offense. Beck remained in prison for five years until July 1901, when he was released for good behavior.

Beck had lived a varied life as a shipping broker, an arms dealer, a copper-mine owner, and even a singer, but now he was broken and lonely. A single man with no family to which to turn, Beck had nothing to dedicate his life to other than doggedly trying to prove he did not swindle the women who accused him and that he was not John Smith. He was unsuccessful. His only accomplishment was to spend what little remained of his fortune, leaving himself with scant resources to contend with his life's next terrible turn.

On April 15, 1904, after three years' freedom had begun to return to him some sense of security, Beck stepped out of his apartment on Tottenham Court Road and onto the street. A young woman, Pauline Scott, ran up to him. "You are the man who took my jewelry and my sovereign," she said. Beck felt as though he had been thrown back into a horrible nightmare with no escape.

"No, I am not," he shouted. "I do not know you; I have never seen you before in my life,"

"You are the man who took my jewelry," Scott shouted back.

Suddenly Beck realized he was not asleep at all and that a conspiracy was afoot. "Who put you up to this?" he demanded. Overwhelmed, Beck started to flee, but he ran straight into the arms of Detective Inspector Ward, who had observed the entire interaction. Ward took him to Paddington police station.

Beck's new arrest had come about after Scott, a maid, had gone to the police complaining that a gray-haired, distinguished-looking gentleman had spoken to her on the street and offered her a position as a housekeeper. The rest of the story followed the pattern of the duping of the women who had accused Beck in 1896.

Inspector Ward, concluding that Beck was back to his supposed old habits, instructed Scott to wait by Beck's door until he came out and then to make her charge. Beck, in his panic, had behaved like a guilty man and tried to run. When the newspa-

Adolf Beck's mugshot

pers reported his arrest, four more women came out of the woodwork claiming to have been similarly swindled. One by one, all of them swore that Beck was the man who cheated them.

Beck appeared once more in the dock at the Old Bailey. He had no funds left with which to hire a seasoned lawyer. "His nose is just the same," Rose Reece, one of the women who came forward, told the jury. "I would pick it out of a thousand." On his own behalf, Beck begged the court to believe that he did not know the women. He believed that he must be the victim of a double. But the trial ended that day and the jury pronounced him guilty. Mr. Justice Grantham, however, felt uneasy about the verdict and postponed Beck's sentencing.

A week later, a man who gave his name as William Thomas was arrested and taken to Tottenham Court Road police station on a charge of hoaxing two unemployed actresses out of their rings. The story of his con coincided in every detail with the crimes for which Beck had been convicted. He looked like Beck, and also had the scar under his ear mentioned by Ottilie Meissonier in her testimony but not found on Beck. Like the description in Smith's prison files and unlike Beck, Thomas was circumcised. It was then that it first began to dawn on police that a terrible mistake had been made. Melville Macnaghten himself took charge of an emergency investigation.

He went to visit the one woman who had refused to testify against Beck in the 1896 trial. At the time, the woman had insisted that a mistake had been made and that Beck was the wrong man, but no one had listened to her. Macnaghten flung down a sheaf of photographs in front of her, and she at once picked up one of Thomas. "That's the scoundrel who robbed me nine years ago, and don't you forget it, Mr. Policeman!"

The five women who had recently identified Beck also viewed Thomas. "My God, that's the man," said Rose Reece.

John Smith's mugshot

The others agreed with her. Those of the witnesses of the 1896 trial who could be reached also viewed the prisoner. They admitted their tragic errors. When at last John Smith's landlord of 1877 recognized William Thomas as his former tenant, the prisoner broke down and confessed to all the crimes attributed to Beck.

The Home Office quickly gave Beck his liberty. On July 19, 1904, he was unconditionally pardoned and given £5,000 compensation (about $450,000 in the year 2000). But the case had stirred up the country. The press and the public denounced Scotland Yard and the Home Office. The government appointed an investigating committee. The case instigated the eventual creation of Britain's first court of criminal appeals. Most importantly, Beck's case shook what police faith remained in the old methods of identification. "The object-lesson in this lamentable business was unquestionably the extreme unreliability of personal identification," wrote Macnaghten in *Days of my Years*.

The entire country agreed with him. The question was how

to avoid such misidentifications in the future. Macnaghten and his boss Edward Henry knew that the consistent use of fingerprints was the answer. To them, the most troubling revelation of the Beck case was the fact that, though scientific evidence might exonerate a man, police still relied on the old methods of personal recognition. Macnaghten and Henry desperately needed a way to convince the stubborn traditionalists to give up their old ways and to embrace scientific identification by fingerprints.

. . .

Eight months after the resolution of the Beck case, Macnaghten awoke on a Monday morning in May 1905 to news of the brutal murders of the frail old couple, Thomas and Ann Farrow, in their Oil and Colour shop in Deptford. Macnaghten rushed to the scene. Blood was everywhere. The public, scandalized, would shout for conviction, yet the vicious burglars had left behind almost no evidence. It was then that, on the emptied cash box under the murdered woman's bed, Macnaghten found the thumbprint of Alfred Stratton, whom police soon arrested along with his brother Albert. Besides the thumbprint, little other evidence linked the Strattons to the crime.

This was the case Macnaghten and Henry needed, one that could burst fingerprinting sensationally upon the public, finally proving the system's power to the judiciary and the police old guard. "Scientific Discovery Nabs Murderers," the headlines might read. But things could also go hideously wrong: "Police Hocus-Pocus Sends Innocents to Jail." Defeat would suggest that Scotland Yard had been willing to risk hanging two innocents for the sake of an experiment. It could mean the end of fingerprints, and probably, of Edward Henry's career. Henry again turned to the crack prosecutor, Richard Muir.

This time, Muir did not hesitate to take the case, but he warned Henry that a jury would not be willing to send men to

the gallows if the newfangled evidence could not be substanti-
ated. No witness interviewed so far had seen the brothers do
anything more incriminating than run down the street. That
was not good enough. Had the brothers been seen on the High
Street in the hours before the murder? Muir wanted to know.
Had they suddenly begun spending money after the murder that
they didn't have before? Henry's and Macnaghten's investigators worked day and night to answer Muir's questions. In their
hands rested the fate of the most important ever trial in the history of scientific identification.

Ten

The Stratton Trial

A hush fell over the Old Bailey courtroom when Richard Muir, adorned in his wig and robes, stood up on May 5, 1905 to open the prosecution against Albert and Alfred Stratton. Crowds of spectators strained on tiptoe to get a glimpse of the famous prosecutor. Reporters scribbled furiously into their notebooks. Alfred and Albert Stratton stared terrified at Muir as though he, and not the judge, might at any moment order their execution.

Muir spoke slowly and deliberately. The vicious way Thomas and Ann Farrow were killed, he told the jury, made this the most brutal case he had ever come across. "The manner in which the faces of the poor old couple had been battered about," he said, "made it quite impossible to extend even the slightest human consideration towards the murderers." He paused.

Muir looked down to read from a pile of paper slips, jokingly referred to as his "playing cards" by the judges and barristers familiar with Muir's methods. In one color pencil, Muir had written out cards for his opening statement, in another, for his chief examination, in another, for the cross. Dangerous points of a case he always marked out in red pencil, the color of the notes he'd made on thumbprint evidence.

Red was also the color in which Muir wrote the name of

Henry Faulds, who sat at the defense table, whispering in the ears of the Strattons' lawyers, preparing to tear Muir's case apart. Faulds had made his objection to the Yard's use of single prints well known. In Faulds's view, proof of single fingerprints' individuality was not sufficiently solid to send a man to his death on their say-so. Nor did he believe that fingerprint identifications should be left in the hands of potentially biased police constables who were not trained scientists.

Faulds's expert advice and testimony in this case could fatally undermine the credibility of the fingerprint system to which he had dedicated his life. But the government and the Yard had made an enemy of Faulds. They had rebuffed and ignored him ever since he first suggested the use of fingerprints, and Faulds's charity had worn thin. He refused to stand idly by while the Yard used the methods Faulds first suggested to wrongly send men to the gallows. Faulds and Melville Macnaghten, fighting on opposites sides of the Farrow case, were both, ironically, concerned with the same thing: preventing the possible conviction of innocents.

Macnaghten and Henry had convinced Richard Muir that Faulds's concerns were misplaced. Nevertheless, Muir tucked his red "playing cards" to the back of the pack. Before presenting the riskiest of his evidence, he would first convince the jury that it was at least reasonable to suspect the Stratton brothers, while playing up the horror of the Farrows' grisly deaths. He began by presenting the circumstantial evidence the police had so carefully collected on Muir's earlier insistence.

During the course of this trial, Muir said, pointing at the Stratton brothers, he would show that Albert and Alfred Stratton had skulked about the vicinity of Chapman's Shop, waiting for the appointed hour of their crime, that they were seen coming from the shop after they'd done their terrible deed, and that they afterwards had more money than they could honestly account for. He would prove that these two men carried iron jemmies with which to force open the door to the shop, but that, finding that the unsuspecting Farrow answered their knocks, they used the tools of their notorious trade instead to callously bludgeon Farrow and his wife to death.

What inextricably connected these local thugs to the crime, Muir went on, were the clues they'd left behind: three stocking masks and a finger-mark on the cash-box tray. Muir would prove that the Stratton brothers were in the habit of using such masks. And evidence, Muir said simply, "would be called for the purpose of satisfying the jury that the finger-mark on the cash-box tray was the mark of Alfred Stratton's right thumb." But that would come later.

First came Muir's march of over forty witnesses through the courtroom. Each witness, by Muir's thinking, dug one spadeful deeper the grave Muir had in mind for the two Strattons. William Jones, the Farrows' young shop assistant, described the terrible shock of discovering Farrow's body. The police surgeon horrified the jury with his descriptions of the Farrows' facial wounds and fractured skulls. Policeman after policeman described the placement of the masks and the cash box at the crime scene, steadily building the importance of this key evidence in the minds of the jury.

Kate Wade, Albert's girlfriend, said he regularly slept all day and stayed out all night. Sarah Tedman, Albert and Kate's one-time landlady, recalled how Albert once brought home a bag full of knives and forks, which Alfred later took away as though to find a buyer. Why was Albert such a night owl? Muir wondered aloud. Why did the brothers traffic in silverware? The jury might conclude, Muir suggested, that the brothers kept the habits of seasoned burglars.

On the Sunday night before the murders, Albert had not stayed with Wade, she testified, nor had he slept at his new lodgings, boardinghouse workers said. Albert was out and unaccounted for on the night of the murders, Muir noted. So where exactly was Albert as the clock ticked its way toward the end of the Farrows' lives? Believing he knew the answer, Muir turned the topic to the stocking masks that, he theorized, tied the brothers to the crime.

Wade testified that Alfred once asked her for stockings. Tedman, the landlady, had found stocking tops, with eye-holes cut out, under Albert's mattress. Muir showed Tedman the masks from the murder scene. Were they like these? Muir asked. They were very similar, Tedman said. The masks under Albert's mattress were like the masks found by the bloodied bodies of Thomas and Ann Farrow? Yes. Muir paused to let this answer sink in, but Tedman had one more nail to hammer into the Strattons' coffin.

"I saw them get something from the top of a chest of drawers in Albert's room," she said, "a long, very bright chisel, and another one with a hook, and a screwdriver." This testimony showed that the brothers owned the heavy, sharp metal tools of a burglar. Muir hastily recalled the police surgeon to the stand. The doctor confirmed that among the tools Tedman described could have been the murder weapon. "The long instrument would, in all probability, have caused the fracture which I found on Farrow," he said. Muir moved confidently on. A quick, grim glance at the jury would say everything that needed to be said.

So far, Muir's evidence had shown that the Strattons had the motive—they were burglars—and the means—they owned the tools—to commit the murders. Now Muir called a number of townspeople to the stand, each placing the Stratton brothers on the Deptford High Street, near the shop, in the hours leading up to the murder. Henry John Littlefield, a professional boxer,

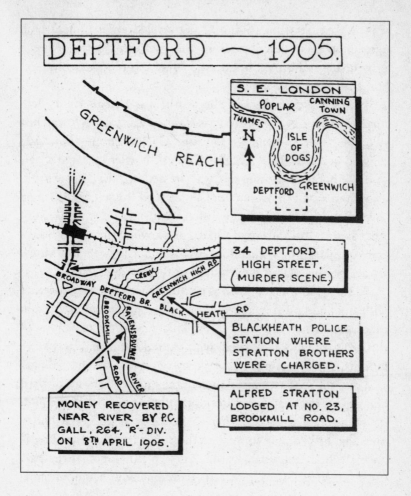

DEPTFORD ~ 1905.

GREENWICH REACH

S. E. LONDON
POPLAR
CANNING TOWN
THAMES
N
ISLE OF DOGS
DEPTFORD
GREENWICH

BROADWAY
CREEK
DEPTFORD BR.
GREENWICH HIGH RD.
BLACK-
HEATH RD
BROOKMILL
RAVENSBOURNE
ROAD
RIVER

34 DEPTFORD HIGH STREET, (MURDER SCENE.)

BLACKHEATH POLICE STATION WHERE STRATTON BROTHERS WERE CHARGED.

ALFRED STRATTON LODGED AT NO. 23, BROOKMILL ROAD.

MONEY RECOVERED NEAR RIVER BY P.C. GALL, 264, "R"- DIV. ON 8TH APRIL 1905.

encountered them at 2:30 A.M. on his way home from a coffee house. An hour later, Mary Amelia Compton met the brothers by the Broadway Theatre on her way home from visiting friends.

Crucially, Ellen Stanton, standing by a chemist's shop on the Deptford High Street, saw two men running from the direction of the Farrows' paint shop at 7:15 A.M., the supposed time of the Farrows' mortal wounds. She recognized one of the men. It was Alfred Stratton. Stanton, Littlefield, and Compton all agreed on their descriptions of the two men. Alfred wore a brown suit and brown boots, and Albert wore a dark-blue suit and a bowler hat.

Just before Stanton saw the brothers, milkman Henry Jennings and his eleven-year-old assistant Edward Russell watched two men exit Chapman's Oil and Colour Shop. "One was dressed in a dark blue serge suit, a dark bowler hat, and a pair of black boots. The other one was dressed in a dark brown suit, a cap and a pair of brown boots," testified Russell. The men he saw perfectly matched the other witnesses' descriptions of Alfred and Albert Stratton. On top of showing that the brothers were burglars who owned tools like the murder weapon, Muir's witnesses now placed the brothers at the murder scene.

Muir's plan was to lock this circumstantial evidence together with his fingerprint evidence. But before he turned the key in that lock, he had one more chain link to add: the testimony of the woman who had lived with Alfred Stratton for twelve months. Annie Cromarty, a twenty-two-year-old charwoman, said that on Sunday, the day before the murders, she and Alfred had too little money to buy food, that he'd gone out in the middle of the night, and that, when she woke up at 9:15 on Monday morning, he had paid to wash in the public baths and for bread, bacon, wood, and coal. In the course of the night of the Farrows' murders, Alfred Stratton had somehow acquired some cash.

In the days after the murders, Annie told the court, Alfred got rid of his brown coat, and he stained his brown boots black. Had he read the descriptions in the papers that said one of the murderers wore a brown coat and brown boots? Muir asked. Alfred had. He had also told Annie to say he was in bed with her on the Sunday night and Monday morning of the murders.

Lastly, Cromarty mentioned, in passing, that Alfred's pants smelled of paraffin on the morning of the murders. Muir jumped up, suddenly excited. Did you ask him why his pants smelled?

"He said, 'I spilled a little over my trousers filling the lamp up the other night,'" answered Cromarty. Alfred was the one who customarily fueled the lamp in their room.

Then, had Alfred ever smelled of paraffin before the morning of the murders?

"I had never noticed that smell."

Then, maybe the smell didn't come from the lamp. Perhaps it came from a paint shop!

Muir had pieced together his evidential puzzle pieces in the jury's mind, he hoped, to make one big picture of guilt. On Sunday night, Alfred Stratton had gone out, returning in the morning smelling like a paint shop and flush with cash he had not had before. He had got rid of or changed the appearance of clothing which may have linked him to the crime, and asked Annie to confirm his alibi. Both brothers were seen in the vicinity of the Farrows' shop in the hours before the murder; men of their description were seen leaving the murder scene. This was the theory Muir had tried to impart.

But as Muir added each new link to his evidential chain, the defendants' barristers immediately lopped it off. Over and over again, the Strattons' lawyers jumped up, artfully questioned Muir's witnesses, and undid what Muir had accomplished. Sarah Tedman, Albert's landlady, admitted that the Strattons had made no effort to conceal the silverware Muir claimed was robber's booty. The brothers had even allowed Tedman to hold a candle for them while they got their tools down from the cupboard. Surely they would have tried to hide the tools if they were evidence of illegal behavior? asked H. G. Rooth, Alfred's barrister.

Could Albert's absence from Carrington House on the night of the murder just mean he slept somewhere else? asked Harold Morris, Albert's barrister. The boardinghouse clerk admitted that Albert had stayed at other lodging houses a number of times. So much for Muir's conclusion that Albert had been out and about on the night of the murders.

As for Jennings and Russell, the milkman and his assistant, neither had picked out Alfred or Albert from a line of men shown to them at Blackheath Road police station. "Looking at the prisoners now," said Jennings, in court, "I am unable to say whether those are the men I saw." The only thing linking the men seen leaving the murder scene with the Stratton brothers was the color of their clothes. How many other pairs of men in Deptford that day had worn blue and brown coats? Ellen Stanton, who said she saw the brothers running away from the Farrows' shop, now told the court, "At Blackheath Road police station I saw a number of men, amongst them being Alfred Stratton. I was not able to pick out the other man who was with him that morning." The prosecutor's main witnesses cannot even identify his suspects! Rooth exclaimed.

Even Muir's star witness, Annie Cromarty, went back on her story when Rooth gave her the chance. Alfred, she told the court, had given his brown coat away to a friend. He had stopped wearing it more than a month earlier. As for his brown boots, he first began blacking them "three, four or five weeks before the murder." It hardly sounded like Alfred was trying to get rid of incriminating evidence. Cromarty also accounted for the money Alfred suddenly came up with. "I am in a family way by him. He told me he had put some money away for me long before the murder."

The next of Muir's evidence to come under fire was his idea that the brothers had been skulking around town on the night of the murder, waiting for an opportunity to commit their crime. Witness after witness admitted that the brothers had not acted at all suspiciously on that night. "Alfred spoke to me first, calling attention to himself," said Mary Compton. "There was no attempt on his part to my knowledge to disguise his being there," said Henry Littlefield. If the brothers were planning a murder, would they not have attempted to hide their presence?

Rooth had taken the teeth out of every bite in Muir's attack. Having done so, he now felt that he could safely put Alfred on the stand to tell his side of the story. What happened that night? Rooth asked him simply. At 2:30 A.M., Alfred explained, my brother tapped on my window. Albert had come to ask me for a loan; he had no money for lodgings. Alfred had no money, either. He told his brother, "You better wait a minute or two and I will slip you in here."

By the time Alfred dressed and opened the door to his lodging house, however, Albert was gone. Alfred walked the short distance into town in search of his brother and found him at the top of Regent Street. This was when they met and greeted Mary Compton and Henry Littlefield. Alfred and Albert then went back to Alfred's boardinghouse. I slept on the bed with Annie, said Alfred, and Albert made himself as comfortable as possible on the floor.

Alfred stepped down from the stand. The so-called evidence against Alfred painted a much prettier picture when Alfred did the drawing. Thanks to Rooth, who had pulled at each tiny thread in Muir's argument, the entire case against Alfred had unraveled. Muir's arguments lay in ruins at Rooth's feet.

The noose, however, would not so easily slip off Albert's head. Muir fired one last evidential cannonball at the younger brother's defense, leaving it in splintered pieces. Muir called to the witness stand William Gittings, an assistant jailer at the Tower Bridge Police Court. Albert and Alfred had been held there one afternoon, in the interval between two sessions of a pre-trial hearing. Gittings had been privy to words Albert might not live long enough to regret.

"How do you think I shall get on?" Albert had asked him. "I do not know," the jailer said. "Is he listening?" asked Albert, meaning his brother in the adjoining cell. Gittings looked in and saw that Alfred was sitting down reading a newspaper. He said to Albert, "No, he is sitting down reading a newspaper." As Gittings told the tale to the court, Albert then said, "I reckon he will get strung up, and I shall get about ten years . . . He has led me into this."

Courtroom spectators gasped as the dramatic trial took yet another turn. Everything in Muir's case against Albert Stratton now rested on the words "He has led me into this." What did they mean? Could they possibly have an innocent explanation? Albert's lawyer did not want to ask his client, at least not on the witness stand where Muir would have a chance to twist Albert's answers around him. Perhaps the jury would consider the statement meaningless. But they might take it as a confession. This is what Muir counted on, and there was nothing Morris could do.

Meanwhile, the judge told the jury that, whatever meaning they gave Gittings's testimony, it could not be used against Alfred. They must disregard it when considering Alfred's guilt. It is not evidence against him, the judge emphasized, leaving Muir with no real evidence yet against Alfred. The fingerprint system of Scotland Yard would have to stand on its own or fall. Muir had nothing to help support its case. He had worked his way through all his "playing cards." The only ones left were those written in red.

. . .

Detective-Inspector Charles Collins stood clutching his sheaf of enlarged photographs, ready to enter the courtroom and mount the witness stand. A blackboard had been wheeled in to allow him to draw diagrams. The jury members sat on the edge of their seats, eagerly awaiting the newfangled evidence Muir had promised would be the crux of this case. Edward Henry and Melville Macnaghten looked anxiously on. The time for the historic presentation of fingerprint evidence in a British murder trial had finally come.

But Richard Muir had had bad news. Dr. John Garson, the esteemed scientist and one-time government advisor on identification, a frequently trusted medical witness at the Old Bailey, and the man Edward Henry had fired and made an enemy of four years earlier, would testify for the defense. Convincing the jury that a single sweat mark could identify a man would already be among the biggest challenges in Muir's career. This made it even worse. Garson's prestige made him an even more dangerous defense witness than Henry Faulds.

Muir called in Detective-Inspector Collins and had him mount the stand. He waited. The courtroom grew silent. Muir looked at the jury, and then at the cash box which, having been so much discussed, had taken on a commanding presence of its own. Muir inhaled deeply. He pointed at the cash box. "There is not the shadow of a doubt that the thumbprint on this money-box, which once belonged to the murdered Mr. Farrow, comes from the right thumb of the defendant Alfred Stratton," he said, finally breaking the silence.

He let his words sink in. If the jury believed what he said was true, then so was his interpretation of every other piece of evidence he had so far presented—the Strattons' premeditation, the stolen money, the vicious beatings. The brothers would go to the gallows, and fingerprinting would likely be embraced throughout Britain. If Muir could not convince the jury of his fingerprint facts, the brothers would be free, fingerprints would have to wait months or years for their wide-spread adoption, and more desperate men like falsely imprisoned Alfred Beck might end up in jail.

Muir studied Collins for a moment, phrasing his first question. How long have you worked with fingerprints? Muir asked. He needed to carefully establish Collins's expert credentials, especially since they now would be subject to comparison with Garson's. "I have been employed in connection with the Finger Print Department since the formation of the finger print system in 1901," Collins told the court. And what has been the main thrust of your work? Collins had been responsible for cataloging the Yard's 90,000-strong collection of fingerprints. Scientific training and high-ranking status Garson had, but the defense lawyers would be hard-pressed to claim he had experience as extensive as Collins's four years of day-in and day-out work with such a huge fingerprint collection.

Muir next had Collins meticulously explain the scientific basis of the fingerprint system. If Muir could make the jury understand the system, then, whatever Garson's quibbles, they could use their own eyes and decide for themselves. Collins demonstrated the apparatus for taking fingerprints, exhibited a sample form on which fingerprints were impressed, and explained how two fingerprints are compared. The final determination of identity, he explained, depended on the correspondence of the minute characteristics—endings, bifurcations, islands—of the fingerprint ridges. Muir quizzed Collins for some time. Only when the details were clear did he address Collins to the present case.

Have you examined that cash box? asked Muir, pointing. "Yes," Collins said. And what did you find? "I found a finger print on the inner tray," he said. Collins pulled out the first of his large photographs to show the jury. The impression was an arch. In parts, the ridges were perfectly clear to the eye, in others, they were smudged, and the lines were difficult to discern, almost as though a dirty liquid had been spilled on the diagram. This was the normal condition of any crime-scene print.

Muir ignored the blurriness. Do you have the fingerprints of the Farrows and the police officers who might have touched the cash box? Collins pulled out several more enlarged photographs, this time of crisp, clear inked fingerprints. The jury could immediately see that their ridges followed much different patterns than those of the cash-box print. "I compared them with the mark upon the cash box," Collins said. And do any of them match? Collins shook his head. "They do not agree."

Collins explained that he had not been able, in fact, to find a match for the print on the cash box, until the day Alfred Stratton was arrested. That day, he took inked prints of Stratton's fingers and compared them to the smudge from the cash box. And what did you find? The eyes of every spectator, jury member, lawyer, and clerk turned toward Collins. This is what everyone had been waiting to see. The journalists ceased scribbling. Henry Faulds stopped whispering to the lawyers. Edward Henry and Melville Macnaghten held their breath.

Only then did Collins finally exhibit an enlarged photograph of the inked impression of Alfred Stratton's right thumb. Instantly, every person in the courtroom could see that, unlike the inked impressions of the Farrows and the policemen, the ridges of Stratton's print clearly followed a pattern similar to the one found on the cash box. The jury looked on, transfixed.

"I have indicated by red lines and figures eleven characteristics in which the two prints agree," Collins said. How many points might be the same in two different fingers? Muir asked. "The highest number of characteristics which we have ever found to agree in the impressions of two different fingers is three." In other words, any two fingerprints with four or more matching characteristics must come from the same finger. Eleven characteristics in agreement were enough to prove guilt nearly three times over, Collins implied.

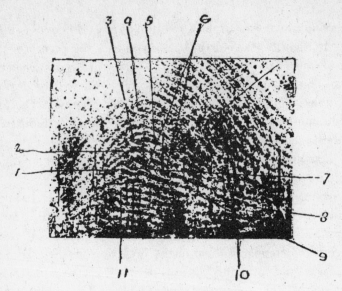

Print from the cash box

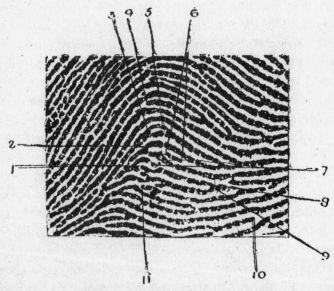

Alfred Stratton's right thumbprint

Collins then described each of the characteristics and pointed out their correspondence to the jury. He showed, for example, how a single ridge coming from the left forked in both prints at point two. At point eleven, in both prints, a ridge coming from the right ended abruptly. Muir waited for him to finish his explanation of all eleven points. Now I want you to think carefully about this, Muir said. Men's lives are at stake. If you have the smallest doubt you must say so. Is it in any way possible that these prints come from different fingers? Collins did not hesitate. "From my experience I should say that it is impossible that those can be prints of two different digits." A murmur went up through the courtroom, and Muir sat down. His case was closed.

Throughout Collins's testimony, Faulds had whispered furiously. Look in the upper right-hand corner of the prints, he told the defense lawyers. The ridges in the cash-box print descend to the right much more steeply than in the inked print. Follow the second and the fifth ridges from that corner in the inked print, he added. They both show splits that cannot be seen in the crime-scene print. The prosecution might say that blurriness in that area made it impossible to know whether the points existed or not. But why, then, could the defense not turn around and say the same of the area that proved the case in their favor? Garson, the first defense witness to mount the stand, would bring up these points in his testimony.

First Rooth got up to cross-examine Collins. He wanted to prove to the jury that Garson was the more senior witness. Questioning Collins, Rooth forced him to admit "I was not exactly a pupil of Dr. Garson" but that "he taught me something to do with finger prints."

Rooth continued to establish the superiority of Dr. Garson's experience when the doctor himself took the stand. Garson answered Rooth's questions succinctly. "I am a Doctor of Medicine." "I have had experience of the finger print system since about 1890." "I was engaged in training a staff of officers in prison and police service." "I had Collins as a pupil in one of my classes." Rooth asked questions about Garson's experience until Rooth felt it was clear that Garson's opinion about the print was more credible than Collins's.

Only then did he ask the question the entire courtroom was waiting for. What is the result of your comparison of the cash-box print and the print of Alfred Stratton's right thumb? For yet another time, the spectators fell quiet in anticipation. "I say they are entirely different fingers," Garson answered. Henry and Macnaghten fidgeted nervously.

Garson then went through Collins's eleven points of comparison one at a time, explaining to the jury why each did not correspond. He based his argument on difference in the curvature of the ridges near the points, and on dissimilarities of the measured distance between them. In different regions, the prints were elongated or compressed compared to each other. Garson discredited every one of Collins's points of "identification," and his testimony infected the courtroom with skepticism. Where Muir thought he had finally turned the key in his evidential lock, Garson had turned it back.

Richard Muir, however, did not intend to give up easily. On cross-examination, he handed Garson a piece of paper. Is that your handwriting? It was. Muir asked Garson to read from it. It was the letter Garson had originally written to the Stratton lawyers offering to testify on the Strattons' behalf. Garson would have no hesitation, the letter said, in testifying "that the way in which [fingerprinting] was being used by the police was just that which would bring it into disrepute."

Muir then produced a copy of a letter Garson had written to the Director of Public Prosecutions. Please also read this out loud for the court. In it, Garson similarly offered to testify for the prosecution. Garson had not suggested in either letter that he would need to analyze the fingerprints in deciding whether they matched or not. He would, it seemed, simply testify for the highest bidder.

What is the date on the letter to the defense? April 26, Garson replied.

And on the letter to the prosecution?

April 26.

"How can you reconcile the writing of those two letters on the same day?" Muir asked.

Dr. Garson shrugged. "I am an independent witness."

"An absolutely untrustworthy one, I should think, after writing two such letters," the judge interjected, devastatingly.

Disgraced, Garson stepped down from the witness stand. Faulds immediately stood up to take his place, but Rooth stopped him. Muir had found a way to destroy Garson's credibility. He probably had a plan to undermine Faulds's as well. Rooth refused to risk further calamity. He closed his case without putting Faulds on the stand. Faulds was crushed. Even here, in the Old Bailey, Henry Faulds was not allowed to take his place as an acknowledged expert on fingerprints.

Muir stood up to make his closing argument. He said he had clearly proved that the Strattons were up to no good on the night of the murders. The fingerprint on the cash box proved conclusively that they had been in the shop and committed the murder. In turn, Rooth argued that there was a perfectly innocent explanation for all the evidence Muir had presented. Muir had tried to sully Garson's reputation, but that didn't change the fact that an eminent scientist who had been a trusted witness in many other trials said the prints were not a match. Was the jury willing to execute a man on this basis?

After receiving their instructions from the judge, who

warned them that they might not want to place too much value on the thumbprint, the jury filed out of the courtroom. In the end, Muir and Rooth had discredited each other's fingerprint testimony. The jury had nothing to rely on but what they had seen with their own eyes.

Eleven

Verdicts

Pacing the Old Bailey's halls while the jury deliberated, Collins, Henry, and Macnaghten were not the only ones concerned about the future of fingerprinting. Their progeny around the world, having nervously kept tabs on the case, now worried about the verdict, too. Hot on the Yard's heels, police in British towns such as Bradford, Blackpool, and York had established fingerprint departments. By 1902, fingerprinting had crossed the Channel to Hungary, Austria, Denmark, Spain, Switzerland, and Germany. In 1903, the New York State Bureau of Prisons became the first to identify United States convicts by fingerprints.

One year later, Scotland Yard sent Detective Inspector John Ferrier to mount a fingerprint exhibition at the World's Fair in St. Louis, Missouri. Inspired by Ferrier's displays, police in St. Louis, Chicago, Oklahoma, and Canada soon adopted the system. Ferrier then visited the newly established National Bureau of Identification in Leavenworth, Kansas, where his demonstrations convinced Federal officials to abandon anthropometric measurements. They created what would eventually become the world's largest fingerprint collection when it was transferred to the Federal Bureau of Investigation.

But by 1905, the year of the Stratton trial, Scotland Yard

and India's Bengal police were the only law enforcement agencies that had used crime-scene prints as evidence in a trial. The Stratton murder case was a worldwide fingerprint experiment, and identification officials everywhere nervously awaited its outcome. A guilty verdict would ease the way for fingerprint departments around the globe to use crime-scene prints in their own important court cases. The system's future was at stake, not just in England, but internationally.

In the prosecution's favor, the fingerprint identification had been made by the world's then top expert, Charles Collins. By that time, Collins's large experience exceeded both that of Faulds, whose daily attention was distracted by his medical practice, and of Garson, who had a far better understanding of anthropometric measurements. Faulds and Garson objected to the identification on the basis of slight differences in the dimensions and ridge flow of the prints from the cash box and Alfred's right thumb. Collins's experience had taught him, however, that such features could never be exactly the same. The fingertip impressed its ridge pattern differently depending on the pressure exerted.

Collins explained this during his cross-examination by Rooth. Can you not see, asked Rooth, that, in the upper right hand corners of the two finger prints, the patterns are entirely different? "I see that the lines in the upper photograph are running much more perpendicular than in the lower, but the pressure on the ridges will alter that," Collins replied. To prove his point, he then took two impressions from a jury member's finger—one pressed lightly, the other hard—and displayed the results. The ridge patterns differed slightly, but the minutia, on which Collins based his identifications, stayed the same.

Was the jury affected more by Collins's or Garson's fingerprint testimony? Did they believe or disbelieve Muir's circumstantial evidence? Exactly what swayed their minds will never be known.

But after only two hours of deliberation, the jury declared that the Strattons were guilty. On May 23, 1905, nineteen days after the trial began, the brothers were hanged. Albert's death was instantaneous, but in Alfred's case, according to the *Times,* "there was some muscular movement after the drop fell."

The case won, the fingerprint profession breathed a collective sigh of relief, and, just as Henry and Macnaghten had hoped, their system gained massive public approval. In the months and years that followed, its use spread rapidly. In 1906, around Britain, police prosecuted four separate cases on fingerprint evidence. The same year, in Birmingham, the system identified its first corpse, the body of a man who had slit his own throat with a knife. Fingerprints would later find widespread application in the identification of disaster victims.

In 1907, the Yard's fingerprint men panicked when, for the first time, a criminal tried to remove his finger ridges. The crook had stepped out of a prison wagon with blood pouring from his fingers and thumbs. He had chopped away at his fingertips with a dirty metal tag attached to his boot lace, hoping to escape identification as a repeat offender. For the few weeks it took his wounds to heal, fingerprint examiners worried that a way to defeat their precious system had been found. But the crook's ridges grew back, his fingerprints were identified, and he received the heavy sentence reserved for habitual criminals.

If Yard officials had made themselves familiar with Faulds's work, they could have predicted this outcome and avoided their panic. Faulds's Japan experiments had shown that finger ridges, whether sliced, rubbed, or burned away with acid, always grew back in their original pattern. However, Scotland Yard was more used to disputing Faulds's views than embracing them, and Faulds still hotly contested their use of single fingerprints to prove identity.

Two years later, in 1909, the law courts settled the single-print argument between Faulds and the Yard, once and for all. In a burglary case, Herbert Castleton had been convicted on the sole evidence of a fingerprint on a candle found at the crime scene. Castleton's lawyers appealed. The appellate court judge asked them: "Can the prisoner find anybody whose fingerprints are exactly like his?" Since the answer was no, the Appeals Court upheld the conviction. Yard officials were overjoyed when, from then on, the precedent of the legitimized use of a single fingerprint as proof of identity in Britain.

Henry Faulds had always insisted that this issue needed confirmation, not in the Appeals Court, but in a scientific laboratory. If single prints were to be used, at the very least Faulds called for the employment of objective scientists whose hands would be tied by the application of stringent evidential standards. Faulds wrote, "It has been laid down that four points of agreement is enough to secure an infallible identification by the 'experts' of Scotland Yard. This is absolute and utter nonsense."

Eventually, official police opinion would come to agree with Faulds. Scotland Yard in 1920 established a fingerprint standard (since revised) that insisted that no crime-scene fingerprint containing less than sixteen points of identification would be sent to court. Sixteen points, the more seasoned fingerprint experts of the time believed, were required to forestall reasonable doubt. That line of thinking was much more in line with the cautious beliefs of Henry Faulds than the more gung-ho standards Charles Collins adhered to in 1905.

Faulds's concern with the misidentification by fingerprints and innocents' possible conviction was part of his motivation for assisting the defense at the Stratton trial. Faulds intended to be the voice of moderation as a foil against the Yard's understandable desire to use the forensic tool wherever possible. Regardless of Faulds's good intentions, the Yard quickly came to regard him as an enemy. Determined to undermine perceptions of Faulds's fingerprint expertise, police officials stood united with William Herschel and Francis Galton in denying his role in the conception of fingerprint identification.

Faulds had not helped his case when, publishing his 1905 *Guide to Finger-Print Identification* soon after the Stratton trial, he publicly denounced the Yard for its use of single fingerprints. He was not without support, however. "Credit where credit is due," proclaimed the *Law Times*, in its favorable review of Faulds's *Guide*. "Mr. Henry Faulds' claims to honourable mention in connection with the system of identification by fingerprints have been strangely overlooked." It said that Faulds's "treatment of the 'mask murders' case is admirable, and a sample of the caution he would enforce upon all investigators when the question is one of life or death."

But the Fingerprint Branch would not go undefended. Out onto the battlefield, to champion the Yard, rode Francis Galton, Faulds's old enemy. Galton wrote an unsigned review of Faulds's *Guide* and published it in the pages of *Nature*. From behind his veil of anonymity, Galton's attack was scathing: "Dr. Faulds in his present volume recapitulates his old grievance with no less bitterness than formerly. He overstates the value of his own work, belittles that of others, and carps at evidence recently given in criminal cases. His book is not only biased and imperfect, but unfortunately it contains nothing new that is of value. . . ."

Four years after writing this review, in 1909, Galton was knighted for his service to science. One year later, Faulds wrote repeatedly to the Home Secretary, Winston Churchill, asking for some similar recognition for his fingerprint contributions. The Home Secretary did not reply.

Faulds made his last desperate plea through his Member of Parliament. On April 19, 1910, the Member stood up in the House of Commons, and asked Winston Churchill whether he had received correspondence from Faulds and what he intended to do about it. Churchill answered: "So far as the Home Office is concerned, I am informed that the adoption of the Finger Print System in 1894 was entirely due to the labours of Mr., now Sir, Francis Galton."

Galton died one year later in 1911, nineteen years before Faulds. But, by Galton's constant insistence, both in print and in private communication, that William Herschel was the sole originator of fingerprinting, Galton had ensured that credit would not come to Henry Faulds until long after he, too, had died. When the credit came, its source was fingerprint examiners from across the Atlantic, where fingerprinting had won its first popular success the year after the Stratton trial.

.   .   .

Towards midnight on April 16, 1906, Detective Sergeant Joseph Faurot of the New York City Police was on patrol by the luxurious Waldorf-Astoria hotel, when he decided to make a quick tour of the Waldorf's corridors to see if the wealthy guests had attracted any thieves. By sheer luck, on the third floor, Faurot came across a British man sneaking out of someone else's suite in stockinged feet. Faurot arrested the Brit, who identified himself as James Jones and insisted that he was a gentleman of the highest social standing.

At police headquarters, protesting that there was a perfectly innocent explanation for his behavior, Jones demanded his release. Faurot's colleagues advised him to accept Jones's explanation and let him go, or risk the disciplinary consequences of the British Consul's potential involvement. But Faurot, on a hunch, charged Jones as a hotel thief, put him in a cell, and sent his fingerprints to Scotland Yard, requesting a check for identification and possible criminal records. If Jones was the gentleman he said he was, Faurot would be in a lot of trouble. Until then, the Brit would have to dine on bread and water while Faurot waited for his reply.

Before being transferred to sidewalk duty, Faurot had worked in the criminal records office at police headquarters, unsuccessfully trying to establish a workable identification system based on anthropometry. In 1904, when word of the Yard's fingerprint success reached New York, Police Commissioner William McAdoo shipped Faurot to London to study the new science. Faurot came home a zealous fingerprint convert, but was not allowed by McAdoo's successor to set up a system. Nevertheless, Faurot's experience at London's Fingerprint Branch led him to send "Jones's" fingerprints to the Yard.

Fourteen days later, the Yard sent word that the prints matched those of Daniel Nolan, a known hotel thief with twelve convictions to his credit, who was wanted for stealing £800 from the house of a famous writer. The Yard's letter included two photographs of Nolan, the spitting image of the prisoner. Faurot had his man and, confronted with the evidence, Nolan admitted his true identity, and was sentenced to seven years in prison. Faurot's fingerprint identification, New York City's first, made a big splash across the front page of the *New York Evening Post.* "Police Learn Lesson from India," the headline proclaimed.

Faurot's second, more important fingerprint victory came in 1908, after the bloody body of Nellie Quinn was found in a rooming house on East 118th Street. Under Quinn's bed, Faurot found a bottle covered with fingerprints that did not belong to the girl. He suspected he might find a match among one of Quinn's "man friends," each of whom Faurot tracked down and fingerprinted, until he came across George Cramer, a plumber. Cramer's prints matched those on the bottle. Confronted with the fingerprint evidence, Cramer confessed that he had killed the girl in a drunken rage.

Fingerprinting had, for the first time in the United States, led to the solution of a murder case. Another two years would pass, however, before fingerprints' evidential use in court to prove guilt in a widely publicized murder case, something the United States needed in order for fingerprinting to become fully accepted. Though the system had spread steadily from state to state since Ferrier's 1904 exhibition, skeptics in police and government still whispered questions about the system's legitimacy. Was fingerprinting truly scientific? Should it be allowed in the courtroom? Any prosecutor winning a conviction on the basis of fingerprint evidence faced the likelihood of an appeal based on these questions—which is exactly what happened in 1911 after the United States' first fingerprint-based murder case was tried against Thomas Jennings.

Clarence Hiller had been fast asleep in his Chicago home when something suddenly woke his wife. She felt frightened, but attributed it to a case of the nighttime jitters. For comfort, she woke her husband, and asked him to light the hallway gas lamp that they usually kept lit, but which had gone out. Hiller's wife watched her husband get out of bed, saw the light splash into the bedroom from the hallway, heard a struggle and then two loud gunshots. Clarence Hiller had been killed by an intruder. When Mrs. Hiller screamed, the intruder fled.

Jennings was arrested later that night when he was found prowling the back streets of Chicago carrying a revolver that had recently been fired. At first, police found little evidence to link Jennings to the murder. Mrs. Hiller could not identify him, and there had been no other witnesses. But Jennings's fingerprints were soon found to match those discovered on a recently painted porch railing at the Hiller home, and he was convicted.

To Jennings's lawyers, his conviction on this newfangled evidence hardly seemed fair. They appealed to the Supreme Court of Illinois, arguing that fingerprint evidence should not be accepted in court until the state legislature had examined the issue and decided whether it passed muster. The question was whether a United States appeals court would follow Britain's lead in accepting fingerprint evidence without such express legislation. It did. In this first test of the legality of fingerprints in an American high court, the landmark ruling stated that "there is a scientific basis for the system of fingerprint identification, and . . . the courts are justified in admitting this class of evidence."

The Illinois Supreme Court's ruling was thorough and comprehensive. It included a complete outline of the history and practice of fingerprint identification. For this reason, proponents of fingerprinting would cite the denial of the Jennings appeal in numerous other fingerprint test cases throughout the United States. The Jennings case gave fingerprinting another boost toward its universal acceptance around the world, but back in Europe, there was still one country where no amount of fingerprint success could convince its identification chief to adopt the system—France.

Since the early days of anthropometry, Bertillon, still the head of the French Identification Service, had gone on to develop many innovative scientific police methods, including the systematic use of crime-scene photography, and the standardized portrait-and-profile mug shot sets still used by police today. He had easily secured his place in police history, but Bertillon remained intensely bitter about the replacement of his identification system by fingerprinting. The man who had once led the world in scientific identification now stalled his department with the technology of the previous century.

Bertillon's one concession to the "Chinese system," as he referred to fingerprints, had been to add space for the inclusion of up to four impressions on his anthropometric cards. If a crime-scene print was found, and the police provided a name,

Bertillon could look up a suspect and compare the crime-scene print with those on his cards. He steadfastly refused, however, to introduce the taking of full fingerprint sets or the classification of identification cards by fingerprints. This meant that, without a suspect's name, Bertillon had no way to cull a match for a set of crime-scene prints from his tens of thousands of files. Fingerprint evidence, therefore, was often useless to him and the French police. This had disastrous consequences for the investigation of one of the country's most famous crimes of its time, the theft of the *Mona Lisa* in 1911.

After the theft, police discovered the painting's abandoned glass case in the museum's service stairway. A nearly complete set of greasy fingerprints marked the glass. Bertillon was called upon to identify the person to whom the prints belonged, but he could do nothing. As a result, police waited two anxious years, until 1913, before the *Mona Lisa* finally turned up. An art dealer helped police trap the robber, Vicenzo Perrugia, after he tried to sell the masterpiece.

Perrugia, it turned out, had been in police custody several times, most recently in 1909 for attempting to steal a prostitute's purse. His fingerprints had been hidden in Bertillon's files the whole time, but Bertillon could not find them because he still classified by measurements. Because of Bertillon's refusal to advance with the times, the recovery of the *Mona Lisa* took a couple of years instead of a couple of hours. The Prefect of Police considered ordering a thorough review of the Identification Service, but, discovering that Bertillon was mortally ill, postponed the action.

The year after the *Mona Lisa* was recovered, in 1914, Alphonse Bertillon died of pernicious anemia. A few weeks later, at an international police conference in Monaco, the main question was how to hunt for international criminals. When the matter of identification arose, Bertillon's successor took the

floor. France's representative at the police conference proclaimed his belief that fingerprints, not anthropometry, should
be adopted as the internationally recognized method of identification. Bertillon's time had passed.

. . .

Around midnight in October, 1917, Sir William Herschel, at age
eighty-four, sat in the study of his home in Berkshire, inking his
fingers and impressing his fingerprints. He was attempting to get
impressions not of the finger ridges, but of the furrows that ran
between them. He did this by lightly impressing his finger until
there was no more ink on the ridges, and then pressing hard so
that the inked furrows came in contact with the paper. He was
still deeply fascinated by the hobby that had become the most
prominent system of criminal identification in the world.

One year earlier, in 1916, prodded by his children, Herschel
published his forty-one-page account of his use of fingerprints in
India, which he presumptuously titled *The Origin of Fingerprinting*. Herschel told the story of how taking the impression
of Konai's hand had led to his official use of fingerprints as signatures in Bengal. He claimed for himself the discovery of fingerprinting, saying that the system stemmed from his giving
Galton a copy of the Konai handprint. Herschel wrote: "The
decisiveness of a finger-print is now one of the most powerful
aids to justice. Our possession of it derives from the impression
of Konai's hand in 1858."

Needless to say, *Origin* did not mention Henry Faulds, and
predictably, its publication began a whole new volley in Herschel's and Faulds's ongoing battle, which had now stretched
over twenty-five years. Having read *Origin*, Faulds returned fire
with a letter written to the editor of *Nature*. Faulds complained
that "Sir William, in his review of the history of this discovery,
has not made any reference to my little contribution on the sub-

ject." He also pointed out that Konai's handprint did not contain the prints of the fingertips, and could hardly, therefore, be
referred to as the source from which fingerprinting was derived.

In sharp terms, Herschel rebuffed Faulds's attack, also in the
pages of *Nature*. He wrote that Faulds's letter "breaks through
all bounds of social courtesy." He then went on to "show reason why I could not honestly have introduced Faulds's name."
According to Herschel, since Faulds had never proved the persistence of finger ridges over many years—something Herschel
believed he himself had proved—Faulds's claim to be the discov-
erer of fingerprinting was therefore false.

In their seventies and eighties, the two old men still scrapped
like stingy schoolboys in a sandbox. Each picked minutely at the
other's claims, until any possibility of reconciling the roles of
their accomplishments was lost. On the night of October 24,
1917, the night that Herschel tried to make impressions of his
fingertip furrows instead of the ridges, he had a seizure and
died. Herschel's niece found him in the morning slumped over
his fingerprints. In a letter she wrote, "He may truly be said to
have carried on his hobby to the very end of his life."

In the 1920s, deprived of his old rival, Faulds no longer
fought his case in the pages of *Nature,* but in the pages of his
own private notebooks. "I have no wish to rob Scotland Yard of
any glory deservedly won, but I do claim that as a faithful citizen I am entitled to some little recognition of the service I ren-
dered," he wrote. He had recently ceased publication of
Dactylography, a fingerprint magazine of which he published
several editions. Having gone deaf, he was forced to close his
medical practice and so could no longer afford to publish the
magazine.

He wrote in his notes, "The only direct reward I have yet
received has been one shilling from Chief Inspector Collins for a
specimen copy of my magazine Dactylography." But money was

Henry Faulds as an old man

not what he wanted. By this time, the roof of his house was near to collapse, and he had nothing of value to leave to his two daughters, but he had never asked Scotland Yard or the government for payment. He did not even want to be called the sole inventor of fingerprints. He merely wanted his name to be footnoted, like Herschel's, below the names of Edward Henry and Francis Galton, who had brought fingerprinting to practical use.

On March 19, 1930, Faulds died, an eighty-six-year-old man still craving recognition for what he considered to be his life's work.

Fifty-seven years later, in 1987, two American fingerprint examiners, researching the history of their profession, stumbled on Faulds's grave at the back of the cemetery of the Parish Church at Wolstanton, Stoke-on-Trent, in the British Midlands. It was unkempt and overgrown. They had searched for it because, starting in 1938, a Scottish judge, George Wilton, began championing Faulds's posthumous case for recognition. Wilton won for Faulds's poverty-stricken daughters two government payments, made on the condition of secrecy. The government continued to refuse to publicly acknowledge a debt to Faulds.

Wilton's campaign may not have moved the British authorities to recognize Faulds, but it did inspire fingerprint examiners around the world to take note of Faulds's role in their profession. Standing in the Wolstanton graveyard, the two American fingerprint experts decided to bear the cost of rehabilitating Faulds's gravesite. The Fingerprint Society, Britain's professional association of fingerprint professionals, committed to paying for the grave's yearly upkeep.

On the new stone, above the emblem of the Fingerprint Society, was inscribed: "In memory of Henry Faulds in recognition of his work as a pioneer in the science of fingerprint identification." Fifty-seven years after his death, the fingerprint profession had finally paid homage to the man to whom they were indebted for the creation of their science.

Epilogue

Though frequent media attention to the identification of criminals by DNA gives the impression that fingerprinting has been left behind in a Sherlock-Holmesian past, the success statistics of the two technologies tell a different story. As of March 2000, the New York City Police Department had used DNA technology to make a grand total of only 200 suspect identifications. In 1999, on the other hand, the force made 1,117 identifications from crime-scene fingerprints in a *single* year, six times the number of identifications made by DNA in its entire NYPD history.

At the federal level, the United States' DNA database receives only about 2,000 requests annually for identification from crime-scene evidence. The FBI's new $164 million Integrated Automatic Fingerprint Identification System (IAFIS), however, chews its way through an annual 85,000 inquiries. These numbers drive home the point that, in the 120 years since Henry Faulds first suggested fingerprints, the importance of their use as a crime-fighting tool has never stopped growing.

In part, this is because the huge leaps in computer technology allow fingerprints to be quickly and easily identified in gigantic databases—IAFIS can hold up to 65 million fingerprint sets. Also, new chemical techniques mean fingerprints can now be lifted off anything from plastic bags to human skin. Most importantly, fingerprints remain as relevant as they ever were because the hand continues to leave behind its calling card on everything it touches. Since the hand is the agent of nearly every human action, everything we do, including our lawbreaking, is autographed in fingerprints' nearly invisible ink.

The idea that criminals generally avoid leaving their fingerprints behind by wiping them away or wearing gloves is a modern myth. Keeping your prints from the police is not as easy as Hollywood makes it seem. How can an outlaw be sure, for example, that he has wiped away what he can't easily see? In 1960, a London rapist closed the shutters of a young mother's bedroom, attacked her, and wiped down only the inside of the shutters with pajamas he found on the floor. He didn't see the place where his thumb had also gripped the outside of the shutters. Confronted with his thumbprint at the Old Bailey, the rapist pled guilty.

Using gloves is equally problematic. Depriving their wearers of touch sense and dexterity, gloves often end up in a criminal's pocket just at the time when they should be on his hands. At crime scenes, fingerprint examiners rush to the surfaces criminals frequently touch barehanded—for example, entrance and exit points, the outdoor pipes robbers sometimes climb up, or the toilet they might have flushed after stripping off their gloves to use it.

Another place examiners know to look is the trash. In 1962, a pair of art thieves heisted works by Matisse, Renoir, Degas, and Picasso from a London art gallery. Smart enough to wear rubber gloves, it never occurred to them that police would look in the trash can, find the package the gloves were sold in, and take fingerprints from it. Even if outlaws never leave a single print on the crime scene, they often don't think to wear their gloves in the getaway car or in their own hideout. After Britain's Great Train Robbery in 1963, a group of professional criminals who would have succeeded in stealing 120 mailbags containing

£2,500,000, left their fingerprints all over their hideout, providing enough evidence for nine of their convictions.

Burglars who get a kick out of making themselves at home in other people's houses make things even easier for police. The robbers stay at the scene much longer than they should, increasing the chance that they'll unconsciously remove their gloves and leave behind their prints. Early in the history of the Yard's Fingerprint Branch, a burglar, during his visit, helped himself to a glass of wine. He even left a sarcastic note of thanks to the owner. The criminal had kicked backed, shed his gloves, and enjoyed a moment of leisure. After all his care not to leave fingerprints anywhere else in the house, he left them on the wineglass.

Keeping the telltale finger autograph away from quick-thinking investigators has become harder as fingerprint technology pushes forward. Though the old-fashioned brush and powder technique, which is still in use, is limited to hard, smooth surfaces, fingerprint examiners can now "lift" a print from just about any surface where a criminal might put his hand. Paper, for example, absorbs the liquid in fingerprints, leaving brushed-on powders nothing to adhere to. Using a technique discovered in 1954, investigators douse paper—a fraudulent bank check, say—in a solution of alcohol and a chemical known as ninhydrin, which stains the absorbed fingerprint sweat patterns and makes them visible.

In another modern technique, discovered in 1982, fingerprint experts waft the heated fumes of superglue over the surface in question. Detectives used this technique in the 1996 investigation of the murder of Martha Hansen, whose naked body was found in the woods of Anchorage, Alaska. Detectives covered Hansen's body with a plastic tent and used an adapted hot air blower to feed superglue fumes into the enclosure. After half an hour, ridge detail, identifying the murderer, showed up on Hansen's left thigh.

Such advances mean that fingerprint identifications can be made in a matter of minutes or hours, at a cost of only pennies. DNA identification, takes anywhere from a couple of days to a couple of months, and can cost thousands of dollars. For these reasons, the use of DNA is still limited to violent crimes. Fingerprints, however, can be used to solve minor crimes such as check fraud and burglary, making them relevant in a much larger number of cases. Henry Faulds's fingerprinting idea is more important than ever.

But nearly a hundred years since the Stratton trial, science has still not provided the one thing Henry Faulds most wanted for his system: proof that no two single fingerprints from different fingers can be the same. The only way to make such a proof would be to collect the fingerprint of every person who ever lived and will live and to mutually compare them. This is impossible. The proof cannot be made.

Just because a proposition—fingerprints are unique—cannot be proved, however, does not mean that it is false. Take the case of another proposition—everyone will die—which is equally unprovable. In 1872, Carl Ernst von Baer, a Prussian embryologist, argued that just because every person in the past had died did not prove that every person in the future would die. Similarly, Faulds worried that just because in the past all fingerprints have been unique does not mean that they will all be unique in the future. In attempting to discredit this logic, fingerprint experts appeal to common sense. No two fingerprints can be the same, they say, just as no two flowers or trees can be exactly alike. The complex set of circumstances that give rise to creation can never be entirely duplicated at two different points in time and space. Each fingerprint is the expression of the biological uniqueness and individuality of each person.

To prove their point, fingerprints' advocates point to identical twins, who are born so physically similar that even DNA testing cannot tell them apart. Yet, though the patterns of the prints on their ten fingers are usually very similar, researchers have always found that the minutia—the ridge endings and bifurcations—remain quite different. Pattern similarity comes from the fact that the arrangement of finger ridges is heavily influenced by genetics. The formation of the ridges, however, is also influenced by the pressures and stresses of growth and the conditions within the womb. It is these factors that cause the differences in the fingerprint minutia of twins.

Nevertheless, there is no proof that two fingerprints will never be found to be the same, just as their is no proof that everyone will die. Based on this reasoning, in his eightieth year, Carl Ernst von Baer declared his belief that some people might not die. In fact, he proclaimed that he might not die.

Perhaps von Baer was right. Perhaps, one day, a man might be born who will live forever. But von Baer died four years after he said he might not. The man who will not die has not yet been found. And today, more than 120 years since Faulds first announced his discovery, neither have we found two people who share the exact same fingerprint.

Source Notes

Because this book is meant as a popular rather than academic account of the history of fingerprints and the role of Henry Faulds, I have not exhaustively annotated my material. I include these source notes, however, in the hope that this might assist future researchers. The names listed below refer to the authors in the bibliography. Throughout the book, I have used quotation marks only where there is a record of the exact words written or said.

ONE
Police investigation into the Farrows murders:
Block, Browne *Fingerprints,* Joseph, Leadbetter, Macnaghten, Stratton, Thorwald, the many 1905 articles listed under "unknown author" from the *Kentish Mercury* and *The Times.*

TWO
Ordeals and the early law of evidence:
Hanson, Lea, Pollock, Stephen.

Murder of Mary Ashford:
Birkett.

Early English law and the Popish Plot:
Bellamy, Hanson, Pollock, Sanderson.

Vidocq and the Sûreté:
Dilnot, *Great Detectives,* Edwards, Griffiths, Thorwald.

The beginnings of British detection:
Browne *Rise,* Dilnot *Great Detectives,* Dilnot *Story,* Emsley, Griffiths.

Early knowledge of fingerprints:
Berry "History," Grew, Lambourne, Wilton *Fingerprints: History.*

THREE

William Herschel and his use of fingerprints in Bengal:
Buttman, Hardcastle "Biographical," Herschel Letter to Bengal, Herschel *Origin,* Herschel Series of his notebooks, Shorland, Spokes Symonds.

Chinese use of fingerprints probable inspiration for Herschel:
Laufer "History," Wilton *Fingerprints: History,* Laufer "Concerning," Wilton *Fingerprints: History.*

Poor quality of fingerprinting at Hooghly Registry under Herschel:
Fingerprint samples, Shreenivas.

History of punishment and criminology:
Cole, Emsley, Foucault, Gould *Mismeasure,* Muncie.

The case of the Tichborne Claimant and its Inspiration of Henry Faulds:
Faulds "Finger Prints: A Chapter," Faulds *Guide,* Stoker, Thorwald *Century,* Wilton *Fingerprints: History,* Griffiths.

FOUR

Conflict between Darwinism and religion:
Gould *Rocks,* Herschel "Skin Furrows," Herschel notes, Kevles, Woodward.

Biography of Henry Faulds:
Berry "Faulds," Checkland, Furuhata, all Faulds's publications, Stewart (Catherine), Stewart (Robert) "Indelible," Stewart (Robert) Personal Interview, Takeuchi, Wilton *Fingerprints: History,* Minutes, Wilton *Fingerprints: History,* Woodward.

Faulds in Japan and his discovery of fingerpints:
Berry "Faulds," Darwin letter to Faulds, Darwin letter to Galton, all Faulds's publications, Furuhata, Galton letter to Darwin, Gould *Mismeasure,* Herschel "Skin," Minutes, Morse, Ritter, Takeuchi, Unknown "Faulds, Henry," Unknown "Omari", Wilton *Fingerprints: History,* Yasoshima.

FIVE
Biographical information on Bertillon and the development of anthropometry:
Bertillon, Chapman "Alphonse," Cole, Rhodes, Sannie, Thorwald.

Faulds's return to England:
Faulds *How*, Wilton *Fingerprints: History*.

Faulds's fingerprint classification system:
Faulds *Dactylography*, Faulds *Guide*, Faulds *Manual*.

SIX
Biography of Galton:
Cole, Forrest, Galton *Memories*, Thorwald.

Galton's studies on heredity:
Cole, Forrest, Galton *Memories*, Galton *Hereditary*, Gould *Mismeasure*.

Galton's and Herschel's collaboration:
Forrest, Galton "Personal Identification," Galton *Memories*, Galton *Finger Prints*, Galton letters to Herschel, Herschel letter to the Editor, Herschel letters to Galton.

Galton's fingerprint studies:
Cole, Forrest, Galton "Personal Identification," Galton *Memories*, Galton "Patterns," Galton *Finger Prints*, Herschel letters to Galton, Thorwald.

SEVEN
Rojas murder and Vucetich:
Chapman "Dr. Juan," Cole, Herroro, Thorwald.

Edward Henry's early career:
"Henry, Edward Richard," Cole, Garvey, Indian Police Collection, Rowland.

Criminal tribes, anthropometry and thumbprints in India:
Cole, Henry *Classification*, Henry reports, Indian Police Collection, Reuben "Khan."

Identification methods in England and the Troup Committee:
Cole, Lambourne, Macnaghten, Thorwald, Troup.

EIGHT
Faulds and Troup:
Faulds *Dactylography,* Faulds *Guide,* Troup, Wilton *Fingerprints History.*

The beginnings of Faulds's feud with Galton and Herschel:
Faulds *Guide,* Faulds *Dactylography,* Forrest, Galton *Finger Prints,* Herschel letter to Bengal Inspector, Herschel letters to Galton, Herschel letter to the editor, Troup, Wilton *Fingerprints: History.*

Development of "Henry Classification System" in India:
Cherrill, Cole, Galton *Finger Print Directories,* Garvey, Henry *Classification,* Henry reports, Henry letters to Galton, Indian Police Collection, Myers "Henry," Reuben "Khan," Shreenivas.

The tea garden murder case:
Henry *Classification,* Shreenivas, Thorwald, Troup, Wilton "Finger-Prints: The Case."

NINE
Lack of success of Troup recommendations in England:
Belper, Cole, Lambourne, Macnaghten, Thorwald.

Adoption of Henry Classification System in England:
Cole, Galton "Identification Offices," Galton letter to Henry, Henry letters to Galton, Garvey, Indian Police Collection, Lambourne, Macnaghten, Reuben "Khan," Shreenivas.

Scotland Yard's early success with fingerprints:
Cole, Block, Browne *Fingerprints,* Felstead, Jackson, Joseph, Lambourne, Macnaghten, Rowland, Thorwald, Unknown "Harry", Wilton *Fingerprints: History.*

Faulds's distrust of single print identifications:
Faulds *Dactylography,* Faulds *Guide,* Faulds *Manual,* Joseph, Unknown "Guildhall", Wilton *Fingerprints: History.*

The case of Adolf Beck:
Thorwald, Unknown "The Case," Watson.

TEN
The Stratton trial:
See references for Chapter One, Felstead.

ELEVEN
Advance of fingerprinting:
Lambourne, Thorwald, Wilton *Fingerprints: History.*

Faulds's view of Stratton case prints:
Berry "Faulds," Faulds *Guide,* Wilton *Fingerprints: History.*

Faulds's last pleas for recognition:
Berry "Faulds," Faulds Notes, Faulds letter to Mitchell, Faulds *The Hidden Hand,* Wilton *Fingerprints: Scotland Yard,* Wilton *Swan Song,* Wilton *Fingerprints: History.*

Faurot cases and Jennings trial:
Illinois vs Jennings, Thorwald, Wilton *Fingerprints: History.*

Bertillon and fingerprints:
Chapman "Alphonse," Rhodes, Sannie, Thorwald.

Herschel's Origin and death:
Hardcastle letter, Herschel *Origin,* Laufer "Concerning," Spokes Symonds, Wilton *Fingerprints: History.*

Faulds's last years:
Faulds Notes, Faulds letter to Mitchell, Berry "Faulds", Wilton *Fingerprints: History.*

EPILOGUE
Comparison of fingerprints and DNA:
NYPD, FBI.

Cases illustrating new fingerprint techniques:
Lambourne, Cummins, Moennsens, Toomey.

Faulds getting recognition:
Berry "Faulds," Wilton *Fingerprints: History.*

Philosophical argument regarding individuality of fingerprints:
Shreenivas, Cummins.

Bibliography

Bellamy, John G. *Criminal Law and Society in Late Medieval and Tudor England*. New York: St. Martin's Press, 1984.

Belper, Lord Henry. "Report of a committee appointed by the secretary of state to inquire into the method of identification of criminals by measurements and finger prints; with minutes of evidence and appendices." London: HMSO, 1901.

Berry, John and Leadbetter, Martin. "The Faulds Legacy." Series of articles in *Fingerprint Whorld*. Oct. 1984 to Jul. 1987.

Berry, John. "History and Development of Fingerprinting." *Advances in Fingerprint Technology*, ed. Lee, Henry C. and Gaesslen, R. E. New York: Elsevier, 1991.

Bertillon, Alphonse. *Signaletic Instructions Including the Theory and Practice of Anthropometrical Identification*. New York: The Werner Company, 1896.

Birkett, Lord. *The New Newgate Calandar*. London: The Folio Society, 1960.

Block, Eugene B. *Fingerprinting: Magic Weapon Against Crime*. New York: McKay, 1969.

Browne, Douglas. *Fingerprints: Fifty Years of Scientific Crime Detection*. London: Harrap, 1953.

Browne, Douglas. *The Rise of Scotland Yard*. New York: G. P. Putnam's Sons, 1956.

Buttman, Gunther. *The Shadow of the Telescope: A Biography of John Herschel* (trans. B. F. J. Pagel). London: Lutterworth Press, 1974.

Chapman, Carey. "Alphonse M. Bertillon: His Life and the Science of Fingerprints." *Journal of Forensic Identification,* June 1993.

———. "Dr. Juan Vucetich: His Contribution to the Science of Fingerprints." *Journal of Forensic Identification,* April 1992.

Checkland, S. G. *Scottish Banking: A History, 1695–1973.* Glasgow: Collins, 1975.

Cherrill, Fred. *Fingerprints Never Lie: Autobiography of Fred Cherrill.* New York: Macmillan, 1954.

Cole, Simon A. "Manufacturing Identity: A History of Criminal Identification Techniques from Photography Through Fingerprinting." Ph.D. dissertation, Cornell University, 1998.

Cotton, H. J. S. Memorandum to H. Holmwood on the subject of the introduction of the scheme of Sir William Herschel for Identification Purposes, The Doepner Collection, International Association of Identification, 16 Jan. 1893.

Cummins, Harold and Midlo, Charles. *Fingerprints, Palms & Soles: An Introduction to Dermatoglyphics.* South Berlin, Mass.: Research Publishing Co., 1943.

Darwin, Charles. Letter to Francis Galton. Apr. 7, 1880. In Lambourne's *The Fingerprint Story.*

———. Letter to Henry Faulds. Apr. 7, 1880. In Wilton's *Fingerprints.*

Dilnot, George. *Great Detectives and Their Methods.* London: Geoffrey Bliss, 1927.

———. *The Story of the Scotland Yard.* London: Geoffrey Bles, 1926.

Edwards, Samuel. *The Vidocq Dossier—The Story of the World's First Detective.* Boston: Houghton Mifflin Company, 1977.

Emsley, Clive. *Crime and Society in England, 1750–1900.* London: Longman, 1987.

Faulds, Henry. Letter to Charles Darwin. Darwin Correspondence Project, MSS Room, University Library, Cambridge. 16 Feb. 1880.

———. "Skin Furrows of the Hand." *Nature.* Vol. XXII. 605. 28 Oct. 1880.

———. "The Beautiful in Nature." *Sunlight.* Vol I. 11. 15 Oct. 1883.

———. *Nine Years in Nipon.* London: Alexander Gardner, 1885.

———. "The Identification of Habitual Criminals by Finger Prints." *Nature.* Vol. L. 548. 4 Oct. 1894.

———. Letter to Francis Galton. Galton Archives, University College London. 3 Dec. 1894.

———. *Guide to Finger-Print Identification.* Hanley: Wood, Mitchell & Co, Ltd., 1905.

———. *How the English Finger-Print Method Arose.* Hanley: Wood, Mitchell and Co., 1905.

————. "Finger Prints: A Chapter in the History of their Use for Personal Identification." *Knowledge*. Vol. XXXIV. 136. Apr. 1911.

————. "Finger Prints: A chapter in their Use." *Scientific American Supplement*. Vol. LXXII. 326. 18 Nov. 1911.

————. *Dactylography*. Halifax: Milner & Co. 1912.

————. "The Permanence of Finger Print Patterns." *Nature*. Vol. XCVIII. 388. 18 Jan. 1917.

————. *The Hidden Hand*. Hanley: Wood, Mitchell & Co., 1919.

————. *Dactylography Journal*. Published in Hanley, Stoke-on-Trent. Vol 1, Nos. 1–3, 1921. Vol 1, Nos. 1–4, 1922.

————. *Manual of Practical Dactylography: A work fo the use of students of the finger-print method of identification*. London: Police Review Publishing Co., 1923.

————. Letter to Ainsworth Mitchell, Doepner Collection, Library of the International Association of Identification. 22 Nov. 1923. Salem, Oregon.

————. *Was Sir E. R. Henry the Originator of the Finger Print System?* (Unknown publisher and date).

————. Handwritten notes. The Doepner Collection. Library of the International Association of Identification. Undated. Salem, Oregon.

Federal Bureau of Investigation. Data provided by public relations office.

Felstead, Sidney Theodore. *Sir Richard Muir: A Memoir of a Public Prosecutor*. London: John Lane, The Bodley Head Limited, 1927.

Fingerprint samples from the Hooghly Registry of Deeds. 172/5A. Galton Archives. University College London, 1878.

Forrest, D. W. *Francis Galton: The Life and Work of a Victorian Genius*. New York: Tiplinger Publishing Co., 1974.

Foucault, Michel. *Discipline and Punish: The Birth of the Prison* (trans. Alan Sheridan). New York: Pantheon Books, 1977.

Furuhata, Tanemoto. "*Shimon no nakatta Keisatsukan* (Policemen Who Did Not Have Fingerprints)," in Bungei Shunjû v. 44, no. 11 (November 1966), pp. 236–246.

Galton, Francis. *Hereditary Genius*. London: Macmillan. 1869.

————. Letter to Charles Darwin. In Lambourne's *The Fingerprint Story*, Apr. 8, 1880.

————. Letter to William Herschel. Galton Archives, University College London. 1 Mar. 1888.

———. "Personal Identification and Description," *Nature,* June 21 and 28, 1888.

———. "Patterns in Thumb and Finger Marks," *Proc of the Royal Society,* November 27, 1890.

———. Letter to William Herschel. Galton Archives, University College London. 28 Nov. 1890.

———. "Method of Indexing Fingermarks," *Nature,* May 28, 1891.

———. "Identification by Finger-Tips," *Nineteenth Century,* Vol. XXX, No. 174. August 1891.

———. *Finger Prints.* (First published in London by Macmillan and Company in 1892). New York: Da Capo Press, 1965.

———. Letter to Edward Henry. Galton Archives, University College London. 5 Oct. 1894.

———. *Finger Print Directories.* London: Macmillan, 1895.

———. Letter to William Herschel. Galton Archives, University College London. 11 Apr. 1895.

———. "Identification Officers in India and Egypt." *The Nineteenth Century,* July, 1900.

———. *Memories of My Life.* London: Methuen & Co., 1908.

Garvey, Maurice. "'Beyond All Doubt, a Truly Great Man': The Life and Times of Sir Edward Henry." *Fingerprint Whorld.* Jan. 1999.

Gould, Stephen Jay. *Rocks of Ages: Science and Religion in the Fullness of Life.* New York: Ballantine, 1999.

———. *The Mismeasure of Man.* New York: Norton, 1981.

Grew, Nehemiah. "The Description and Use of the Pores in the Skin of the Hands and Feet." Philosophical Transactions of the Royal Society of London, 1684.

Griffiths, Arthur. *Mysteries of Police and Crime.* London: Cassell and Company.

Hanson, Elizabeth. "Torture and Truth in Renaissance England." *Representations,* Spring 1991: 53–82.

Hardcastle, S. W. Letter to Herman Doepner. Doepner Collection. Library of the International Association of Identification. 22 Jan. 1932. Salem, Oregon.

———. "Biographical Sketch of Sir W. J. Herschel Bt." Unpublished manuscript. The Doepner Collection, Library of the International Association of Identification. 1923.

Henry, Edward. "Report from E R Henry Inspector-General of Police to Chief Secretary of the Gov of Bengal." Galton Archives, University College London. 1 May, 1894.

———. Letter to Francis Galton. Galton Archives, University College London. 2 Oct. 1894.

———. Letter to Francis Galton. Galton Archives, University College London. 4 Oct. 1894.

———. Letter to Francis Galton. Galton Archives, University College London. 6 Oct. 1894

———. Letter to Francis Galton. Galton Archives, University College London. 11 Nov. 1894.

———. Letter to Francis Galton. Galton Archives, University College London. 28 Nov. 1894.

———. Letter to Francis Galton. Galton Archives, University College London. 17 Nov. 1899.

———. Letter to Francis Galton. Galton Archives, University College London. undated. 1899.

———. Letter to Francis Galton. Galton Archives, University College London. 24 June 1900.

———. Letter to Francis Galton. Galton Archives, University College London. 20 June 1900.

———. Letter to Francis Galton. Galton Archives, University College London. 5 Oct. 1902.

———. "Report from E. R. Henry to the Chief Secretary to the Government of Bengal." Galton Archives, University College London: 24 June 1896.

———. Letter to Herschel. Galton Archives, University College London. 3 Sept. 1896.

———. Classification and Uses of Finger Prints. London: G. Routledge, 1900.

———. "Memorandum on the Working of the Fingerprint System of Identification: 1901–4." London: New Scotland Yard, 1904.

"Henry, Edward Richard." The India Office List for 1895. London: Harrison and Sons, 1895.

Herrero, Antonio. "Fifty Years of Dactyloscopy in Argentina." Finger Print Magazine. Oct. 1943.

Herschel, William J. Handwritten notes and other materials. History of fingerprints collection. Bancroft Library. University of California, Berkeley. c1870–1910.

———. "Skin Furrows of the Hand." *Nature*. Vol. XXIII. 76. 25 Nov. 1880.

———. Letter to Francis Galton. Galton Archives, University College London. 2 Mar. 1888.

———. Letter to Francis Galton. Galton Archives, University College London. 7 May 1888.

———. Letter to Francis Galton. Galton Archives, University College London. 2 Jul. 1888.

———. Letter to Francis Galton. Galton Archives, University College London. 15 Jul. 1890.

———. Letter to Francis Galton. Galton Archives, University College London. 12 Sept. 1890.

———. Letter to Francis Galton. Galton Archives, University College London. 8 Sept. 1890.

———. Letter to Francis Galton. Galton Archives, University College London. 13 Oct. 1890.

———. Letter to Francis Galton. Galton Archives, University College London. 7 Nov. 1890.

———. Letter to Francis Galton. Galton Archives, University College London. 11 Nov. 1890.

———. Letter to Francis Galton. Galton Archives, University College London. 26 Nov. 1890.

———. Letter to the Editor. *The Times*. Unknown date. 1909.

———. *The Origin of Finger-Printing*. Oxford: H. Millford, 1916.

———. Letter to Bengal Inspector-General of Gaols. *Fingerprints: History, Law and Romance*. George Wilton. Glasgow: William Hodge. 1938.

Illinois vs. Jennings, Supreme Court of Illinois, 252 Illinois 534. 21 Dec. 1911.

Indian Police Collection. MS Eur F161/185. Oriental and Indian Collection. British Library, London.

Jackson, Harry. 11 Central Criminal Court Session Papers 976 (1902).

Joseph, Anne M. "Anthropometry, the Police Expert, and the Deptford Murders: The Contested Introduction of Fingerprinting for The Identification of Criminals in Late Victorian and Edwardian Britain." *Docu-*

menting *Individual Identity: The Development of State Practices Since the French Revolution*, eds. Jane Caplan & John Torpey. Princeton: Princeton University Press.

Kevles, Bettyann. "Charles Darwin." *Encyclopedia Brittanica* (online edition). http://www.britannica.com/bcom/eb/article/5/0,5716,117775+1,00.html

Lambourne, Gerald. *The Fingerprint Story.* London: Harrap. 1984.

Laufer, B. "Concerning the History of Finger-Prints." *Science.* 25 May 1917.

Laufer, Berthold. History of the Finger Print System. *Annual Report of 1912 of the Smithsonian Institution.*

Lea, Henry C. *Superstition and Force: Essays on the wager of law, the wager of battle, the ordeal, torture.* New York: Greenwood Press Publishers, 1968 (Originally printed by Henry C. Lea in 1870.)

Leadbetter, Martin. "Rex v. Stratton and Stratton." *Fingerprint Whorld.* Jan. 1977.

Macnaghten, Melville. *Days of My Years.* London: Edward Arnold. 1915.

Minutes of the Church of Scotland Foreign Mission Committee, National Library of Scotland, Edinburgh.

Minutes of the United Presbyterian Church Foreign Mission Committee, National Library of Scotland, Edinburgh.

Moenssens, Andre. *Fingerprint Techniques.* Radnor, Penn.: Chilton Book Co., 1971.

Morse, Edward S. *The Shell Mounds of Omori.* Tokyo: The University of Tokio, 1879.

Muncie, J., E. McLaughlin, and M. Langan (eds). *Criminological perspectives: A Reader.* London: Sage Publications. 1996.

Myers, Harry J. "History of Identification in the United States." *Finger Print Magazine,* Oct. 1938.

———. "The Henry System Semi-Centennial." *Finger Print Magazine,* June 1950.

New York City Police, data provided by public relations office.

Pollock, Sir Frederick and Maitland, Frederic William. *The History of English Law Before the Time of Edward I.* 2nd Ed. Cambridge: Cambridge University Press, 1968. (Originally published 1895.)

Reuben, Ezekiel. "Khan Bahadur Azizul Haque." *Fingerprint Whorld.* Apr. 1984.

Rhodes, R. T. F. *Alphonse Bertillon: Father of Scientific Detection.* London: Harrap. 1956.

Ritter, H. *A History of Protestant Missions in Japan.* Tokyo: The Methodist Publishing House. 1898.

Rowland, John. *The Fingerprint Man: The Story of Sir Edward Henry.* London: Whitefriars Press, Ltd., 1959.

Sanderson, Edgar. *Judicial Crimes: A record of some famous trials in English history in which bigotry, popular panic, and political rancour played a leading part.* London: Hutchinson & Co., 1902.

Sannie, Dr. Charles. "Alphonse Bertillon and Finger Print Identification." *Finger Print Magazine.* Feb. 1951.

Shorland, Eileen. "Sir William James Herschel and the Birth of Fingerprint Identification." *Library Chronicle of the University of Texas at Austin*, 1980.

Shreenivas and Sinha, Sardindu Narayan, "Personal Identification by the Dermotoglyphics and the E-V Methods." *Patna Journal of Medicine*, Vol XXXI, No. 2. Feb 1957.

Spokes Symonds, Ann. "Herschel's Retirement years." *Fingerprint Whorld.* Oct. 1990.

Stephen, Sir James Fitzjames. *A History of the Criminal Law of England.* London: Macmillan and Co., 1883.

Stewart, Catherine. Personal Interview. Oxted. June, 1999.

Stewart, Robert E. "Indelible Proof of a Pioneer." *The Scottish Field*, Sept. 1992.

Stewart, Robert E. Personal Interview. Edinburgh. June, 1999.

Stoker, Bram. *Famous Imposters.* New York: Sturgis & Walton Company, 1910.

Stratton (Alfred and Albert Earnest), 142 Central Criminal Court Session Papers 978 (1905).

Takeuchi, Hiroshi, ed. Rainichi Seiyô jinmei Jiten. 1983.

Thorwald, Jorgen. *Century of the Detective,* trans. Richard Winston and Clara Winston. New York: Harcourt, Brace & World, 1965.

Toomey, Sheila. "A Forensics Feat; Palm Print Lifted Off Victim's Skin is a First For Alaska, A Rarity Anywhere." *Anchorage Daily News,* 21 Oct. 1996.

Troup, Edward. "Report of a Committee Appointed by the Secretary of State to Inquire into the Best Means Available for Identifying Habitual

Criminals with minutes of evidence and appendices." Personal library of John Berry. 12 Feb, 1894.

Unknown author. "The Metropolitan Police—Identification." *The Times*. Unknown date. 1909.

Unknown author. "Faulds, Henry." *Kodansha Encyclopedia of Japan*. Tokyo; New York: Kodansha, 1983.

Unknown author. "Harry Jackson, 42, labourer . . ." *The Daily Telegraph*. 15 Sept. 1902.

Unknown author. "Morse, Edward Sylvester (1838–1925)." *Kodansha Encyclopedia of Japan*. Tokyo; New York: Kodansha, 1983.

Unknown author. "Omori Shell Mounds." *Kodansha Encyclopedia of Japan*. Tokyo; New York: Kodansha, 1983.

Unknown author. "At the Guildhall yesterday. . . ." *The Times*, 26 Sept. 1902.

Unknown author. "Central Criminal Court." *The Times*, 6 May 1905: 19.

Unknown author. "Central Criminal Court—The Deptford Murder Trial." *The Times*, 8 May 1905: 4.

Unknown author. "Double Execution." *The Times*, 24 May 1905: 7.

Unknown author. "Inquest." *The Times*, 31 Mar. 1905: 11.

Unknown author. "Inquest." *The Times*, 20 Apr. 1905: 8.

Unknown author. "Inquest." *The Times*, 21 Apr. 1905: 8.

Unknown author. "Murder by Masked Burglars." *The Times*, 28 Mar. 1905: 11.

Unknown author. "Murder Most Foul." *Kentish Mercury*, 12 May 1905.

Unknown author. "Shocking Tragedy at Deptford." *Kentish Mercury*, 31 Mar. 1905.

Unknown author. "The Case of Mr. Adolf Beck." *The Times*, 19 Oct., 20 Oct., 21 Oct., 24 Oct., 25 Oct. 1904.

Unknown author. "The Deptford Murder." *The Times*, 1 Apr. 1905: 10.

Unknown author. "The Deptford Murders." *The Times*, 4 Apr. 1905: 14.

Unknown author. "The Murder by Masked Men at Deptford." *Kentish Mercury*, 7 Apr. 1905.

Unknown author. "The Murder by Masked Men at Deptford." *Kentish Mercury*, 14 Apr. 1905.

Unknown author. "The Murder by Masked Men at Deptford." *Kentish Mercury*, 21 Apr. 1905.

Unknown author. "The Murder by Masked Men at Deptford." *Kentish Mercury,* 28 Apr. 1905.

Unknown author. "The Murder by Masked Men at Deptford." *Kentish Mercury,* 5 May 1905.

Unknown author. Report on the British Association meeting. *The Times.* 15 Sept. 1899.

Watson, Eric R. *Adolf Beck.* Edinburgh and London: William Hodge & Co., 1924.

Wilson, G. Fleetwood. "Report of the Committee on Identification by Finger Prints." Personal library of John Berry, 1902.

Wilton, George W. "Finger-Prints: The Case of Kangali Charan, 1898." *The Juridical Review,* 1 Dec. 1937.

Wilton, George W. *Fingerprints: History, Law and Romance.* Glasgow: William Hodge. 1938.

———. *Fingerprints: Scotland Yard and Henry Faulds.* Edinburgh: W. Green and Son, Limited, 1951.

———. *Fingerprints: The Swan Song of Old "Dr Fingerprints."* North Berwick, East Lothian, Scotland: Tantallon Press. 1963.

Woodward, Sir Llewellyn. *The Age of Reform: 1815–1870.* Oxford History of England, vol. 13. Oxford: Clarendon Press, 1962.

Yasoshima, Shinnosuke. "Henry Faulds—Pioneer in Dactyloscopy His Life and Work in Japan." *Fingerprint and Identification Magazine,* May 1960.

Index